MCQ's on Entomology

About the Authors

I. Yimjenjang Longkumer is currently pursuing his Ph.D programme in Dr. Rajendra Prasad Central Agricultural University, Pusa, Bihar under the guidance of Dr. A. K. Misra, Professor, Department of Entomology, P.G. College of Agriculture, RPCAU. He had completed his B.Sc. (Agril.) from School of Agricultural Science and Rural Development, Nagaland and M.Sc. (Agril.) from College of Agriculture, Central Agricultural University, Manipur. He has published research papers and reviews related to insect pest management of various crops.

Dr. Ashok Kumar Misra is currently working as University Professor in the Department of Entomology. P.G. College of Agriculture & Director Students Welfare and different positions in Dr. Rajendra Prasad Central Agricultural University, Pusa, Bihar. He completed his Ph.D from Banaras Hindu University, Varanasi. He has a wide experience of more than 25 years of Teaching, Research and Extension experience in the research area of Integrated Pest Management, Biological control and crop pests etc. and working as Co-ordinator for AICRP on Rice at RPCAU Pusa, Centre. To his credit, he has a number of popular articles and research publications in very reputed journals in India & Abroad.

Dr. Md. Abbas Ahmad is presently working as Assistant Professor-cum-Scientist in the Department of Entomology, P.G. College of Agriculture, Dr. Rajendra Prasad Central Agricultural University, Pusa (Bihar). He obtained his Doctoral degree from G.B. Pant University of Agriculture and Technology, Pantnagar. He has more than 12 years experience in Teaching and Research. He has several National and International publications. He is now continuing his research on All India Co-ordinated Research Project on Pigeon pea and Chickpea as Co-PI (Entomology).

MCQ's on Entomology

I. Yimjenjang Longkumer
Ashok K. Misra
Md Abbas Ahmad

books for everyone
nipa PAPERBACKS
New Delhi – 110 034

nipa books for everyone
PAPERBACKS

NEW INDIA PUBLISHING AGENCY

101, Vikas Surya Plaza, CU Block, LSC Market
Pitam Pura, New Delhi – 110 034, India
Email: info@nipabooks.com
Web: www.nipabooks.com

For customer assistance, please contact
Phone: + 91-11-27 34 17 17 Fax: + 91-11- 27 34 16 16
E-Mail: feedbacks@nipabooks.com

ISBN 978-93-89130-24-9

Composed and Designed by NIPA.

Preface

Entomology is a branch of Zoology which deals with the study of Insect and it is derived from Greek word "Entomon" which means Insect and "logia" means knowledge. Insect constitutes the highest number of living species. They have the ability to reproduce in high numbers within short time, highly specialized morphological features, capacity to fly and their survival in different adverse ecologies makes a constant wonder and enables them to dominate on earth.

This book envisages the basic information pertaining to the insect's anatomy, physiology and morphology. It is imperative to have a constructive knowledge about the biological process, factors enhancing the process occurring rapidly inside their body system. This ultimately helps us to have an in-depth information about their behaviour in the ecosystem which includes interaction with their host, life cycle etc.

We hope that this book will provide a comprehensive view to all the inspiring readers preparing for competitive exams such as JRF/SRF/ARS/State Entrance exam and other competitive exams related to Agricultural Entomology.

I.Y. Longkumer
A.K. Misra
Md. Abbas Ahmad

Contents

Acronym

AICRPBC & W	All India Coordinated Research Project on Biological Control of Crop Pests and Weeds (Bangalore)	1977
BPPD	Biopesticides and Pollution Prevention Division (USA)	1967
CAL	Central Agmark Laboratory (Nagpur, Maharashtra)	1986
CIL	Central Insecticide Laboratory (Faridabad, Haryana)	1971
CSB	Central Silk Board (Bangalore)	1948
CSRTI	Central Sericultural Research & Training Institute (Berhampore, Mysore etc.)	1961
DPPQS	Directorate of Plant Protection Quarantine and Storage (Faridabad, Haryana)	1946
EPA	Environment Protection Act (USA)	1986
ESI	Entomological Society of India (New Delhi)	1938
FDCA	Food Drug and Cosmetic Act (USA)	1938
FEPCA	Federal Environmental Pesticide Control Act (USA)	1972
FIA	Federal Insecticide Act (USA)	1910
FIFRA	Federal Insecticide Fungicide Rodenticide Act (USA)	1947
FQPA	Food Quality Protection Act (USA)	1996
FSS	Food Safety and Standards Act (New Delhi)	2006
ICZN	International Commission on Zoological Nomenclature (Singapore)	1895
IIT	Indian Institute of Toxicology (Pune, Lucknow)	1965
IPFT	Institute of Pesticide Formulation Technology (Gurgaon, Haryana)	1991
IRAC	Insecticide Resistance Action Committee	1984
NBAIR	National Bureau of Agricultural Insect Resources (Bangalore)	1993
NCIPM	National Research Centre for Integrated Pest Management (New Delhi)	1988
NFSA	National Food Security Act (India)	2013
NIHSAD	National Institute of High Security Animal Diseases (Bhopal, Madhya Pradesh)	2014
NIPHM	National Institute of Plant Health Management (Hyderabad, Telangana)	1966
NIVEDI	National Institute of Veterinary Epidemiology and Disease Informatics (Bangalore, Karnataka)	2013
NPPTI	National Plant Protection Training Institute (Hyderabad, Telangana)	1966
NSSO	National Silkworm Seed Organization (Bangalore)	1975
OPCW	Organization for Prohibition of Chemical Weapons (The Hague, Netherlands)	1997
PCAI	Pest Control Association of India, Chennai	1976

PFA	Pure Food Act (USA)	1906
PFA	Prevention of Food Adulteration Act (India)	1954
PMFAI	Pesticide Manufacturers and Formulation Association of India (Mumbai, Maharashtra)	1967
RPTL	Regional Pesticides Testing Lab (Kanpur and Chandigarh)	1985

1

Insect Morphology

1. Insect heart is located
   a) Dorsally
   b) Ventrally
   c) Pleurally
   d) Dorsal lateral
2. Shell glands or Green glands are the excretory organ of
   a) Crustacean
   b) Onychophora
   c) Myriapoda
   d) Chelicerata
3. Entognathous mouth parts are found in insect order
   a) Diplura
   b) Collembolan
   c) Protura
   d) All of the above
4. The head is formed by the fusion of
   a) 4 embryonic segments
   b) 5 embryonic segments
   c) 6 embryonic segments
   d) 7 embryonic segments
5. Insect wings are located
   a) Ventro-laterally on the 2nd and 3rd thorasic segment
   b) Dorso-laterally on the 2nd and 3rd thorasic segment
   c) Dorso-ventrally on the 2nd and 3rd thorasic segment
   d) All of the above
6. Which part of the insect integument is the seat of complex biochemical changes
   a) Epidermis
   b) Cuticle
   c) Basement membrane
   d) Procuticle
7. The non-cellular layer of the insect integument is
   a) Procuticle
   b) Epicuticle
   c) Cuticle
   d) Cuticulin
8. The first layer to be secreted in the cycle of cuticle formation is
   a) Epidermis
   b) Cement layer
   c) Lipid layer
   d) Cuticulin

9. The order of cuticle deposition is
   a) Cuticulin layer – wax layer – procuticle – cement layer
   b) Cuticulin layer – procuticle – wax layer – cement layer
   c) Cuticulin layer – procuticle – cement layer
   d) Cuticulin layer-wax layer – Cement layer

10. Procuticle consist of
   a) Endocuticle
   b) Exocuticle
   c) Cuticle
   d) Both a and b

11. Chitin is insoluble in water, alkalis, dilute acids and organic solvents but soluble in
   a) Sodium hypochlorite
   b) Concentrated mineral acids
   c) Both a and b
   d) None of the above

12. The adjacent epidermal cells of insect integuments are held with one another by means of
   a) Endoplasmic reticulum
   b) Golgi apparatus
   c) Desmosomes
   d) Mitochondria

13. An imago state of an insect immediately after molting whereby the body is soft and immature in coloring
   a) Prepupa
   b) Propupa
   c) Teneral
   d) Instar

14. The hydrofuge property of the cuticle surrounding the spiracles is due to the presence of
   a) Rectal glands
   b) Siphons
   c) Peristigmatic glands
   d) Rectal pappillae

15. Arthropodin is an untanned protein, water soluble and is the predominant composition of
   a) Endocuticle
   b) Exocuticle
   c) Epicuticle
   d) Cuticulin

16. Sclerotin is a tanned protein, water insoluble and is the predominant composition of
   a) Endocuticle
   b) Exocuticle
   c) Epicuticle
   d) Cuticulin

17. Cutice is divided into two parts namely
    a) Cement layer + lipid layer
    b) Cuticulin + wax layer
    c) Endocuticle + Exocuticle
    d) Procuticle + Epicuticle
18. Procuticle has two parts namely
    a) Cement layer + lipid layer
    b) Cuticulin + wax layer
    c) Endocuticle + Exocuticle
    d) Procuticle + Epicuticle
19. The zone of deposition of new cuticle or Procuticle also called as Schmidts layer lies between
    a) Endocuticle and epidermis
    b) Epidermis and basement membrane
    c) Procuticle and Epicuticle
    d) Cement layer and lipid layer
20. Chitin is made up of chains of
    a) N-acetylglucosamine and glucosamine units
    b) α-acetylglucosamine and glucosamine units
    c) β-acetylglucosamine and glucosamine units
    d) γ-acetylglucosamine and glucosamine units
21. The chains of α-acetylglucosamine and glucosamine units are linked by
    a) 1-1 beta linkage
    b) 1-2 beta linkage
    c) 1-3 beta linkage
    d) 1-4 beta linkage
22. The protein responsible for the flexibility of wings is
    a) Arthropodin
    b) Sclerotin
    c) Resilin
    d) All of the above
23. The chitinous part of cuticle is
    a) Procuticle
    b) Epicuticle
    c) Both a and b
    d) None of the above
24. Quinones which are responsible for tanning and Sclerotization in insect cuticle are derived from
    a) Protease
    b) Chitinase
    c) Tyrosine
    d) Hydroquinone
25. Verson's gland which is present in Lepidoptera is responsible of secreting which part of epicuticle
    a) Cement layer
    b) Wax layer
    c) Cuticulin layer
    d) Polyphenol layer

26. Cement layer is also known as
    a) Hypodermis b) Hyperdermis
    c) Tectocuticle d) Exodermis
27. The process of conversion of Arthropodin into Sclerotin is
    a) Sclerotization b) Tanning
    c) Hardening d) Both a and b
28. The epidermal cells are anchored to the basement membrane by
    a) Desmosomes b) Hemidesmosomes
    c) Endoplasmic reticulum d) Golgi apparatus
29. The non-cellular appendages of the integument are
    a) Spines b) Ridges
    c) Microtrichia d) All of the above
30. Multicellular movable cuticular appendages is
    a) Spines b) Spurs
    c) Bristles d) Setae
31. A comb-like structure called ctenidia is prominent in
    a) Siphonaptera b) Coleoptera
    c) Diptera d) Hymenoptera
32. Setae are hollow structures developed as extensions of the
    a) Endocuticle b) Exocuticle
    c) Procuticle d) Epicuticle
33. Setae (Macrotrichia) arises from a cup-like pit called
    a) Aculei b) Alveolus
    c) Tormogen cell d) Hollow pit
34. Microtrichia are also known as
    a) Aculei b) Alveolus
    c) Tormogen cell d) Trichogen cell
35. Setae is produced by a single, enlarged cell called as
    a) Tormogen cell b) Trichogen cell
    c) Schwann's cell d) Retinular cell
36. Lepidopteran families characterized by the presence of stinging setae includes
    a) Limacodidae b) Arctiidae
    c) Saturinidae d) All of the above

37. The arrangement of bristles on the insect based on the size and position is called

a) Chaetotaxy b) Phonotaxis
c) Rheotaxis d) Thigmotaxis

38. The outgrowths which are integral part of insect integument having no membranous articulation are

a) Cuticular appendages b) Cuticular processes
c) Cuticular joints d) Cuticular hairs

39. The outgrowths which are non-cellular in origin and lacks basal articulation is

a) Cuticular appendages b) Cuticular processes
c) Cuticular joints d) Cuticular hairs

40. An infolded part of the walls of the thorax of an insect is

a) Tagma b) Tapetum
c) Septum d) Phragma

41. The outer non cellular layer of body wall is

a) Cuticulin b) Cuticula
c) Procuticle d) Endocuticle

42. Hollow invagination of the body wall is known as

a) Apodeme b) Apophysis
c) Holodeme d) Hypodeme

43. Solid invagination of the body wall is known as

a) Apodeme b) Apophysis
c) Holodeme d) Hypodeme

44. Detachment of the cuticle from the epidermis is called

a) Apodeme b) Apophysis
c) Apolysis d) Holodeme

45. The lines or grooves which divides the surface of the head into regions is called

a) Suture b) Sclerites
c) Tagma d) Tagmosis

46. The area between the sutures is called

a) Suture b) Sclerites
c) Tagma d) Tagmosis

47. The endoskeleton of insect head is called

a) Apodemes
b) Phragmata
c) Apophysis
d) Tentorium

48. Melanization is the process where cuticle becomes

a) Dark
b) Hard
c) Light
d) Soft

49. Sclerotization is the process where cuticle becomes

a) Dark
b) Hard
c) Light
d) Soft

50. Tyndall blue coloration is observed in insect order

a) Odonata
b) Dictyoptera
c) Ephemeroptera
d) Dermaptera

51. Structural colouration occurs in

a) Lampyridae
b) Buprestidae
c) Phengodidae
d) Apionidae

52. Scoli (Thorny process) is a characteristic features of

a) Nymphalid larvae
b) Saturniid larvae
c) Both a and b
d) Embids larvae

53. The body segments of an insect are grouped together to form three usually well-defined regions called

a) Tagmosis
b) Phragma
c) Tagmata
d) Phragmosis

54. The differentiation of insect body into functional regions is called

a) Phragmosis
b) Tagmosis
c) Metameres
d) Somites

55. The dorsal and facial surface of the head of immature insects is crossed by ecdysial cleavage line having the form of inverted Y shape, the ecdysial cleavage line is also referred to as

a) Epistomal suture
b) Frontoclypeal suture
c) Epicranial suture
d) Fronto genal suture

56. The snout in the weevils are formed by fusion of

a) Fron and Gena
b) Fron and Clypeus
c) Fron and vertex
d) Fron and vertex

57. The fron bears the median ocellus and is delimited by

a) Frontogenal sulci laterally
b) Epistomal sulcus
c) Frontoclypeal sulcus
d) Clypeal suture

58. Antennary segment is innervated by

a) Protocerebrum
b) Deutocerebrum
c) Tritocerebrum
d) Sub-oesophageal ganglion

59. Labrum is innervated by

a) Protocerebrum
b) Deutocerebrum
c) Tritocerebrum
d) Sub-oesophageal ganglion

60. The area from the occipital to post occipital suture is

a) Post occiput
b) Occiput (U shaped)
c) Genae
d) Clypeus

61. The mouth parts of Siphunculata (Anoplura) order is

a) Rasping type
b) Piercing and sucking type
c) Sponging type
d) Chewing type

62. The mouthparts of Psocoptera and Mallophaga is

a) Mandibulate (Chewing) type
b) Piercing type
c) Sucking type
d) Sponging type

63. The suture between Clypeus and Fron is

a) Occipital suture
b) Post occipital suture
c) Epistomal suture (Clypeo-frontal suture)
d) Ocular sutures

64. The only true suture having metameric significance is

a) Occipital suture
b) Post occipital suture
c) Epistomal suture (Clypeo-frontal Suture)
d) Ocular sutures

65. The facial area of the cranium just above labrum in which the dilator muscles of the cibarium are attached on its inner surface is

a) Fron
b) Gena
c) Clypeus
d) Vertex

66. The portion of the epicranium which lies behind the fron and between the compound eyes is
   a) Gena b) Subgena
   c) Vertex d) Occiput

67. Posterior opening of the cranium is
   a) Epicranium b) Post occiput
   c) Epistomal d) Occipital foramen

68. The median ventral sclerite present in Coleoptera is called
   a) Mala b) Gula
   c) Ligula d) Glossa

69. The type of insect head in which long axis of the head is vertical and mouth parts are ventrally place and projected downwards is
   a) Prognathous b) Hypognathous
   c) Opisthognathous d) Hypernathous

70. The type of insect head in which the long axis of the head is horizontal and mouthparts are directed forward is
   a) Prognathous b) Hypognathous
   c) Opisthognathous d) Hypernathous

71. The type of insect head in which the head is directed towards backwards so that the mouthparts are directed backwards and arises between the anterior legs is
   a) Prognathous b) Hypognathous
   c) Opisthognathous d) Hypernathous

72. Pilifer/labral pilifer a hearing organ is found on the heads of
   a) Apionidae b) Curculionidae
   c) Chrysomelidae d) Sphingidae

73. Johnston's organ is absent in insect order
   a) Collembolan and Diplura b) Thysanura and Embioptera
   c) Collembola and Thysanura d) Diplura and Thysanura

74. The semicircular plate above the antennal bases and below the ptilinal fissure called lunula is present in
   a) Chironomidae b) Asilidae
   c) Mycetophilidae d) Tephretidae

75. The antennae is adapted for seizing prey in larvae of
   a) Chaoborus b) Meloe
   c) Chalcids d) Braconids

76. Which of the following insect family are typically characterized by extremely long antennae
   a) Bruchidae b) Meloidae
   c) Cerambycidae d) Cucujidae

77. The sub order Nematocera of Diptera are characterized by their
   a) Long, segmented antenna b) Short antenna
   c) Clavate antenna d) Geniculate antenna

78. The sub order Brachycera of Diptera are characterized by their
   a) Long, segmented antenna b) Short antenna
   c) Clavate antenna d) Geniculate antenna

79. Which is the first segment of antennae
   a) Scape b) Pedicle
   c) Flagellum d) Funicle

80. Microscopic sensilla in the form of flattened sense organs on the antennae of aphids is termed
   a) Sensilla trichoidea b) Rhinaria
   c) Sensilla placodea d) Sensilla Campaniform

81. The suborder of Diptera characterized by long, segmented antenna is
   a) Nematocera b) Brachycera
   c) Cyclorrhapha d) All of the above

82. The suborder of Diptera characterized by short, segmented antenna is
   a) Nematocera b) Brachycera
   c) Cyclorrhapha d) All of the above

83. The mandible of Pterygotc insects is a solid compact piece articulating with the head by
   a) Ginglymus b) Condyle
   c) Both a and b d) Abductor muscles

84. The flexible plate fringed with hairs present in the inner body of mandibles is
   a) Dendricles b) Prothesca
   c) Stipe d) Cardo

85. The condition in which mandibles are protrusible, rotatory structures enclosed with the maxillae in a gnathal pouch that is sunk into the head is

a) Ectognathous b) Entognathous
c) Exognathous d) Epignathous

86. The primary jaws or true jaws of insect mouthparts is

a) Maxillae b) Mandilbles
c) Labrum d) Labium

87. The stylets of female mosquitoes are composed of

a) Labrum-epipharynx b) Labium-epipharynx
c) Labium d) Labrum

88. Stridulation in family Reduviidae are produced by

a) Rostrum and protergum b) Forewings and Hindwings
c) Rostrum and prosternum d) Forelegs and forewings

89. In chewing and lapping mouthparts the glossa terminates in to a spoon shaped lobe called

a) Haustellum b) Flabellum
c) Rostrum d) Labellate

90. Which one of the following mouth part is absent in rasping and sucking type

a) Right mandible b) Left mandible
c) Right maxillae d) Left maxillae

91. Which of the following insect family have distinctive mandibles found on the males of most species, which resemble the antlers of stags?

a) Anobiidae b) Bostrichidae
c) Elateridae d) Lucanidae

92. The type of mouthparts in bed bugs is

a) Chewing and Biting type b) Siphoning type
c) Piercing and sucking type d) Chewing and Lapping type

93. Mandibles are absent in which type of insect mouth part

a) Chewing and Biting type b) Siphoning type
c) Sponging type d) Chewing and Lapping type

94. A pair of dorsolateral lobes arising from the hypopharynx of an insect

a) Lingua b) Superlingua
c) Dendricles d) Stipes

95. The condition when the mouthparts are enclosed within the head capsule is

a) Entognathous b) Ectognathous

c) Monocondyl d) Dicondyl

96. The grub of antlion has which type of mouthpart

a) Chewing and lapping b) Rasping and sucking

c) Mandibulosuctorial type d) Sponging type

97. In sponging type of mouthpart the end of the proboscis is sponge like and has two lobe and acts as suction pads called

a) Flabellum b) Haustellum

c) Labella d) Rostrum

98. The distal part of the proboscis in sponging type of mouthpart adapted for sucking the exposed liquids is

a) Flabellum b) Haustellum

c) Labella d) Rostrum

99. In siphoning type of mouthpart the proboscis is formed by

a) Labium b) Labrum

c) Galea of maxillae d) Mandibles

100. The portion of the maxillae that is attached to the head is

a) Cardo b) Stipes

c) Galea d) Lacinia

101. Palpifer is present in

a) Maxillae b) Mandible

c) Labium d) Labrum

102. Palpiger is present in

a) Maxillae b) Mandibles

c) Labrum d) Labium

103. Palpiger is borne on

a) Submentum b) Prementum

c) Mentum d) Stipes

104. The two lobes of maxillae are

a) Cardo and stipes b) Subgalea and galea

c) Lacinia and galea d) Lacinia and subgalea

105. In many coleopteran larva each maxilla carries a single lobe formed by galea and Lacinia called

a) Gula b) Lula
c) Mala d) Both a and b

106. The mouthparts of Lepidopteran insects are siphoning type except in the family

a) Pyralidae b) Papilionidae
c) Nymphalidae d) Micropterigidae

107. The structure formed by the fusion of mandibular plates with Hypopharynx is

a) Gula b) Lura
c) Mala d) Phragma

108. In labium the outer pair (paraglossae) and an inner pair (glossae) together constitute

a) Tegula b) Ligula
c) Labial Palp d) Submentum

109. Which part of the mouth in dragonfly nymph is modified as mask

a) Labrum b) Maxillae
c) Labium d) Mandibles

110. Mandible attached with two hinges to the head capsule is referred to as

a) Dicondyl b) Mocondyl
c) Prothesca d) Both a and b

111. Mandible attached with two hinges to the head capsule is referred to as

a) Dicondyl b) Mocondyl
c) Prothesca d) Both a and b

112. Osmeteria which is an repugnatorial gland located in prothorasic segment occurs in

a) Papilionidae b) Lycaenidae
c) Hepialidae d) Noctuidae

113. The sides of the pronotum in lepidoptera consists of a lobe like structure known as

a) Torneus b) Cuneus
c) Patagia d) Clavus

114. The pleuron of the pterygotes consists of

a) Episternum (anterior sclerire) b) Epimeron (posterior sclerite)
c) Pleural sulcus d) All of the above

115. The functional base of insect leg is

a) Coxa b) Trochanter
c) Femur d) Tibia

116. The sequence of the insect leg part is

a) Coxa, trochanter, femur, tibia and tarsus
b) Coxa, femur, trochanter, tibia and tarsus
c) Coxa, femur, trochanter, tarsus and tibia
d) Coxa, tarsus, trochanter, femur and tibia

117. The trochanter is divided into two subsegments called (Trochantellus) in insect order

a) Odonata and parasitic Hymenoptera
b) Odonata and Coleoptera
c) Odonata and Embioptera
d) Odonata and Dictyoptera

118. Plantulae is a part of

a) Leg b) Mouth
c) Wings d) Thorax

119. The median dorsal process at the end of tarsus to which the Pretarsus are articulated is

a) Unguifer b) Alula
c) Gaster d) Metastoma

120. In which family of insect are physically characterized by having hydrofuge hairpiles, retractable preapical claws and elongated legs and body

a) Nepidae b) Coreidae
c) Miridae d) Gerridae

121. Pleuston insect belongs to family

a) Nepidae b) Coreidae
c) Miridae d) Gerridae

122. Pretarsus consists of a pair of claws and a medium lobe called

a) Arolium b) Empodium
c) Pulvilli d) Plantulae

123. In aristate type of insect leg, a median spine like structure is represented by
   a) Arolium b) Empodium
   c) Pulvilli d) Plantulae
124. Prolegs/Pseudolegs are characteristic features of
   a) Larvae of Lepidoptera b) Larvae of Coleoptera
   c) Larvae of Diptera d) Larvae of Hymenoptera
125. The tip of proleg is called
   a) Planta b) Crochets
   c) Hooks d) Claspers
126. The raptorial type of legs in mantids are modified
   a) Fore legs b) Mid legs
   c) Hind legs d) Both b and c
127. Crochets are absent in the larva of
   a) Monotrysia b) Symphyta
   c) Both a and b d) Ditrysia
128. The forelegs are greatly reduced in which of the following Lepidopteran family
   a) Noctuidae b) Hepialidae
   c) Nymphalidae d) Palaeosetidae
129. The two lobes lying below the claws is called
   a) Pulvilli b) Plantulae
   c) Empodium d) Arolium
130. An opaque spot present near the costal margin of the insect wing is
   a) Pterostigma b) Cuneus
   c) Embolium d) Clavus
131. The anterior region of insect wing containing bulk of veins in front of the claval furrow is called
   a) Vannus b) Cunues
   c) Tornues d) Remigium
132. The posterior area of the insect wing is flexible and is termed
   a) Vannus b) Cunues
   c) Tornues d) Remigium

133. An ergatoid is a

a) Wingless reproductive adult ant b) Winged reproductive adult ant

c) Workers of ant d) Soldiers of ant

134. Twisted wing (Hind wings at rest) insects belongs to order

a) Diptera b) Trichoptera

c) Strepsiptera d) Megaloptera

135. Which of the following insect group, wings are completely absent

a) Female Embioptera b) Grylloblattodea

c) Female Psychidae d) All of the above

136. In some insects the posterior region of the wing base consists of a pair of lobes known as

a) Calypter b) Alula

c) Anal fold d) Both a and b

137. The leading edge of the wing is called

a) Costal margin b) Apical margin

c) Anal margin d) Humeral margin

138. Insect with characteristic braided wings belongs to order

a) Diptera b) Mecoptera

c) Siphunculata d) Plecoptera

139. The trailing edge of the wing is called

a) Costal margin b) Apical margin

c) Anal margin d) Humeral margin

140. The angle at the base of the wing is called

a) Costal angle b) Anal angle

c) Apical angle d) Humeral angle

141. The angle between the costal and apical margin is

a) Apical angle b) Anal angle

c) Humeral angle d) Costal angle

142. A small lobe of the base of wings that rest on hind wings found in Lepidopteran insects is

a) Fibula b) Jugum

c) Tonues d) Vannus

143. The hind-wings and fore-wings are of equal length in
   a) Diptera b) Lepidoptera
   c) Isoptera d) Zoraptera
144. Mechanism of direct flight is seen in
   a) Mayflies b) Dragonflies
   c) Damselflies d) All of the above
145. The angle between the apical and anal margin is
   a) Apical angle b) Anal angle
   c) Humeral angle d) Costal angle
146. A pair of small scale like sclerites at the base of costa of each fore wing is
   a) Pterostigma b) Embolium
   c) Tegulae (paraptera) d) Axillary sclerites
147. Sound produced by rubbing hind leg against the adjacent forewing is seen in
   a) Grasshopper b) Crickets
   c) Katydids d) Longhorned beetles
148. Sound produced by rubbing one wing against the other wing is seen in
   a) Grasshopper b) Crickets
   c) Katydids d) Both b and c
149. Sound produced by the back edge of the pronotum against a file on the mesonotum occurs in
   a) Grasshopper b) Crickets
   c) Katydids d) Longhorned beetles
150. Which of the moths family often bear metallic scales
   a) Gelechiidae b) Yponomeutidae
   c) Choreutidae d) Plutellidae
151. The sclerite by which the wings of an insect articulates to the thorax is
   a) Pteralia b) Basilar sclerites
   c) Subalar sclerites d) All of the above
152. Four axillary sclerites are present in
   a) Orthoptera and hymenoptera b) Orthoptera and lepidoptera
   c) Orthoptera and Coleoptera d) Hymenoptera and lepidoptera

153. Flexing mechanism of wing is absent in insect order
   a) Ephemeroptera and Odonata b) Coleoptera and Hymenoptera
   c) Lepidoptera and Coleoptera d) Hymenoptera and lepidoptera
154. The flight movements of an insect is caused by
   a) Direct muscles (Largest) b) Indirect muscles
   c) Accessory indirect muscles d) All of the above
155. Degenerate trachea known as Semper's rib is present in the wings of
   a) Coleoptera b) Orthoptera
   c) Lepidoptera d) Hemiptera
156. On the base of halters two large sensory bodies are present which are termed
   a) Hick's papillae b) Rick's papillae
   c) Semper's papillae d) Schwan's papillae
157. Alula is present in
   a) Coleoptera b) Lepidoptera
   c) Hymenoptera d) Diptera
158. During histological changes of insect wings, blood filled spaces which soon develop in the wing pad is
   a) Lacunae b) Tracheae
   c) Scale socket d) Plasma
159. The hindwings of Dipteran insects are modified into
   a) Halteres b) Pseudohalteres
   c) Tegmina d) Elytra
160. Tegula is present in insect order
   a) Lepidoptera b) Diptera
   c) Hymenoptera d) All of the above
161. The forewings of Strepsiteran insects are modified into
   a) Halteres b) Pseudohalteres
   c) Tegmina d) Elytra
162. Fringed wings occurs in
   a) Zoraptera b) Mallophaga
   c) Thysanura d) Thysanoptera

163. Primitive wing venation is seen in
   a) Bruchidae b) Aleyrodidae
   c) Micropterigidae d) Meloidae
164. The thoracic segment consist of dorsal sclerite called
   a) Pleuron b) Sternum
   c) Pronotum d) Notum
165. The notum of prothorax is called
   a) Pronotum b) Pleuron
   c) Sternum d) Tergum
166. The veins which follows the ridges are termed
   a) Convex veins (+) b) Concave veins (-)
   c) Biconcave d) Biconvex
167. The veins which follows the furrows are termed
   a) Convex veins (+) b) Concave veins (-)
   c) Biconcave d) Biconvex
168. The hypothetical wing venation consists of
   a) Costa, subcostal, radius, radial sector, media, cubitus and anal veins
   b) Costa, radius, subcostal, radial sector, media, cubitus and anal veins
   c) Costa, subcostal, radial sector, radius, media, cubitus and anal veins
   d) Costa, subcostal, radius, radial sector, cubitus, media and anal veins
169. A whitish spot called thyridium which has a sensory function is present on the fore wings of
   a) Trichoptera b) Hymenoptera
   c) Neuroptera d) Raphidioptera
170. The presence of Pterostigma in dragonflies and damselflies are also seen in
   a) Snakeflies (Raphidioptera) b) Hymenopterans
   c) Megalopterans d) All of the above
171. A scar-like area called thyridium on each side of tergite which lies between the anterior margin and the spiracle is found in
   a) Ichneumonidae b) Tenthrinidae
   c) Trichogrammatidae d) Platygasteridae

172. Spurious vein (thick vein) is present in
   a) Braconidae b) Syrpidae
   c) Formicidae d) Apidae

173. Trigamma (Fusion of median vein, Cu1a and Cu1b) is the characteristic branching of veins found in
   a) Forewings of Hymenoptera b) Forewings of Dermaptera
   c) Forewings of Lepidoptera d) Forewings of Coleoptera

174. The scented scales relcased from the wings of males butterflies is
   a) Meconium b) Androconia
   c) Stigma d) Onchonia

175. Wingless sterile females in a colony of ants are
   a) Workers b) Soldiers
   c) Drones d) Both a and b

176. Adult Protura and Diplura bears a pair of appendages on the first abdominal segment known as
   a) Collophore b) Pleuropodia
   c) Papillae d) Pseudopodia

177. In Apocrita suborder the narrow waist formed between the first two segments of the abdomen is called
   a) Propodeum b) Petiole
   c) Gaster d) Pedicle

178. The plump portion of the abdomen posterior to the petiole is called
   a) Propodeum b) Petiole
   c) Gaster/Metasoma d) Pedicle

179. The wax secreted by the worker bees for constructing their honeycomb is located in
   a) $3^{rd} - 6^{th}$ abdominal segments b) $3^{rd} - 7^{th}$ abdominal segments
   c) $4^{rd} - 7^{th}$ abdominal segments d) $2^{nd} - 6^{th}$ abdominal segments

180. Styliform appendages is common in
   a) Endopterygote b) Exopterygota
   c) Apterygota d) All of the above

181. Collembolans possess three pregenital appendages known as Collophore, tenaculum and furcula which are located on
   a) 1st, 4th, 3rd abdominal segments respectively
   b) 3rd, 4rd and 1st respectively
   c) 1st, 3rd and 4th respectively
   d) 3rd, 4th and 1st respectively

182. The leaping organ on the ventral side of the abdomen in collembola is
   a) Retinaculum b) Tenaculum
   c) Furcula d) Frenulum
183. Presence of pair of forceps-like pincers on the abdomen occurs in
   a) Isoptera b) Zoraptera
   c) Dermaptera d) Psocoptera
184. Females without wings, mouths and legs belongs to
   a) Papilionidae b) Psychidae
   c) Pieridae d) Nymphalidae
185. Asymmetrical anal cerci is present in
   a) Male Embioptera b) Female Embioptera
   c) Male Dermaptera d) Female Dermaptera
186. Dorso-ventrally flattened insect includes
   a) Dermaptera b) Anoplura (Siphunculata)
   c) Cimicidae d) All of the above
187. Laterally compressed body insects belongs to
   a) Siphunculata b) Siphonaptera
   c) Raphidioptera d) Megaloptera
188. Sternal spatula or breast bone is a characteristic feature of family
   a) Cecidomyiidae b) Tipulidae
   c) Mycetophilidae d) Chironomidae
189. In male Odonata the copulatory organ is developed on the ventral surface of the
   a) 1st and 2nd abdominal segments b) 2nd and 3rd abdominal segments
   c) 3rd and 4th abdominal segments d) 4th and 5th abdominal segments
190. The lateral clasping organ in male genitalia is called
   a) Aedagus b) Parameres
   c) Ovipositor d) Valvulae
191. In the females of insect the ovipositor consist of
   a) Valvifers (3 pairs) and Valvulae (3 pairs)
   b) Valvifers (2 pairs) and Valvulae (3 pairs)
   c) Valvifers (1 pairs) and Valvulae (3 pairs)
   d) Valvifers (3 pairs) and Valvulae (3 pairs)

192. Valvifers are the appendages present in the 8th and 9th abdominal segment which functions as
   a) Basal sclerite for muscle attachment
   b) Apical sclerite for aiding in oviposition
   c) Tactile receptors
   d) Mechano receptors

193. Valvulae are the appendages present in the 8th and 9th abdominal segment which functions as
   a) Basal sclerite for muscle attachment
   b) Apical sclerite for aiding in oviposition
   c) Tactile receptors
   d) Mechano receptors

194. Most aphids have a pair of cornicles (siphunculi), abdominal tubes on the dorsal surface of their
   a) Second abdominal segment
   b) Third abdominal segment
   c) Fourth abdominal segment
   d) Fifth abdominal segment

195. The defensive droplets released by the cornicle is
   a) Triacylglycerols
   b) Terpenoids
   c) Flavonoids
   d) Phenolics

196. The structure producing a violent "click" that enables the click beetle to bounce into the air and back to its original position includes
   a) Prosternum
   b) Mesosternum
   c) Metasternum
   d) Both a and b

197. Which type of abdomen in Coleoptera the hind coxae is fused with metathorax and 1st, 2nd and 3rd abdominal sternite are more or less fused
   a) Adephagid
   b) Haplogastrous
   c) Symphiogastrous
   d) Hologastrous

198. Which type of abdomen in Coleoptera the 2nd abdominal sternite represents a pleurite and a small lateral plate represents sternite
   a) Adephagid
   b) Haplogastrous
   c) Symphiogastrous
   d) Hologastrous

199. The 2nd abdominal sternite distinct to 3rd sternite is a type of which abdomen in Coleoptera
   a) Adephagid
   b) Haplogastrous
   c) Symphiogastrous
   d) Hologastrous

200. In which type of abdomen in Coleoptera the pleurite of the 2nd abdominal segment is fused to that of the 3rd and the sternite is membranous

a) Adephagid b) Haplogastrous

c) Symphiogastrous d) Hologastrous

201. Honeydew secreted from the lepidopteran family is

a) Noctuidae b) Pyralidae

c) Lycaenidae d) Micropterigidae

202. The suborder of Thysanoptera having a tube shape like abdomen is

a) Tubulifera b) Terebrantia

c) Symphyta d) Both a and b

203. The suborder Terebrantia of Thysanoptera order have

a) Pseudo ovipositor b) Round ovipositor

c) Conical or Saw like ovipositor d) Needle like ovipositor

204. In hymenoptera the first abdominal segment is fused with meta thorax which is termed

a) Propodeum b) Hypodeum

c) Hyperdeum d) Holodeum

205. The pygidial glands present in sub order insect of adephagan is located

a) Posterodorsally in the abdomen b) Ventrally

c) Laterally d) Anteriodorsally

206. Sensory hairs called trichobothria present on the abdominal segments occurs in

a) Pyrrhocoridae b) Reduviidae

c) Gyrinidae d) Hydrophilidae

207. The chemical compounds ejected by the bombardier beetles as a defense mechanism are

a) Hydroquinone b) Hydrogen peroxide

c) Cantharidin d) Both a and b

208. A pair of glandular organs (embryonic appendages) located on the first abdominal segment of an insect is

a) Furcula b) Pleuropodia

c) Collophore d) Retinaculum

209. The condition in which the abdomen of the queen termite becomes bloated after mating due to enlargement of ovaries is known as

a) Physiogastry
b) Phagocytosis
c) Pinocytosis
d) Physogastry

210. Dolichaster is a segmental protuberance fringed with setae and are present in

a) Larvae of *Chrysopa*
b) Larvac of Mosquito
c) Larvae of Sawfly
d) Larvae of Antlion

211. The Cornicles/Siphonculi are located dorsally on the

a) 1st or 2nd abdominal segment
b) 2nd or 3rd abdominal segment
c) 4th or 5th abdominal segment
d) 5th or 6th abdominal segment

212. Ovipositor is modified into sting in

a) Diptera
b) Hymenoptera
c) Lepidoptera
d) Hemiptera

213. The dorsal surface of the caudal segment of the abdomen consist of a vasiform orifice which function to aid in the secretion of honeydew is a characteristic features of

a) Aphid
b) Whitefly
c) Scale insects
d) Mites

214. Within the orifice a tongue like structure is present which is known as

a) Superlingua
b) Ligula
c) Lingula
d) Prothesca

215. Microscopic structure like granules called brochosomes are secreted by insects belonging to family

a) Cicadellidae
b) Psyllidae
c) Fulgoridae
d) Cicadidae

216. Caudal breathing tube is present in

a) Tingidae
b) Dysticus
c) Dragongly
d) Water scorpion

217. Pseudo ovipositors are present in

a) Bees and Ants
b) Wasps and Ants
c) Hornets and Bees
d) Fruitflies and Houseflies

218. A pair of outgrowths of the Tergum of segment 9 in the form of short spines or multiarticula is known as

a) Cerci b) Urogomphi

c) Ovipositor d) Sting

219. The Tergum of the 11th abdominal segment is represented by

a) Paraproct b) Gonophysis

c) Gonoxite d) Epiproct

220. A pair of lateral sclerites present in the apex of abdomen is known as

a) Paraproct b) Gonophysis

c) Gonoxite d) Epiproct

221. Pygostyles is a pair of unsegmented cerci like structures found in

a) Braconids b) Earwigs

c) Aphids d) Scoliid wasp

222. The function of anal cerci is

a) Tactile receptors b) Mechano receptors

c) Gustatory d) Olfactory

223. Network like lace in pronotum and forewings occurs in

a) Tingidae b) Reduviidae

c) Polyctenidae d) Dytiscidae

224. Median caudal filament is present in

a) Thysanura b) Ephemeroptera

c) Protura d) Both a and b.

225. Urogomphi is present in the 9th abdominal segment of

a) Diptera b) Odonata

c) Coleoptera d) Dermaptera

226. Dufour's glands are located in

a) Females of Apocrita b) Males of Apocrita

c) Females of Symphyta d) Males of Symphyta

227. The presence of which compounds in the milkweed bugs protect them from predation

a) Cardenolides b) HCN

c) Phenolics d) Quinones

228. In which division of Heteroneura (infraorder) belonging to order Lepidoptera the female has two distinct sexual openings: one for mating, and the other for laying eggs

a) Monotrysia
b) Glossata
c) Aglossata
d) Ditrysia

2

Insect Nervous System

229. An insect's nervous system is a network of specialized cells called
   a) Neuropile b) Dendrites
   c) Axon d) Neurons
230. The afferent neurons are also called as
   a) Sensory neurons b) Motor neurons
   c) Internuncial neurons d) Association neurons
231. The efferent neurons are also called as
   a) Sensory neurons b) Motor neurons
   c) Internuncial neurons d) Association neurons
232. Which of the following neurons carry information towards the Central Nervous System
   a) Sensory neurons b) Motor neurons
   c) Internuncial neurons d) Association neurons
233. Which of the following neurons conduct signals away from the CNS and stimulate response in muscles and glands
   a) Sensory neurons b) Motor neurons
   c) Internuncial neurons d) Association neurons
234. The neurons of CNS in insect is mostly
   a) Bipolar b) Unipolar
   c) Multipolar d) Tetrapolar
235. Afferent neurons are mostly
   a) Bipolar b) Unipolar
   c) Multipolar d) Both a and c
236. Efferent neurons are mostly
   a) Bipolar b) Unipolar
   c) Multipolar d) Both a and c

237. Individual nerve cells are connected to one another through a junction called

a) Synaptis b) Synapsis

c) Synopsis d) Cynasis

238. Group of neurons are called

a) Ganglion b) Nerve sheath

c) Ganglia d) Neuropile

239. Which of the following regulate swallowing movement in insect

a) Ingluvial ganglia b) Frontal ganglia

c) Commissures d) Glial cells

240. Which of the following regulate peristaltic movement in insect

a) Ingluvial ganglia b) Frontal ganglia

c) Commissures d) Glial cells

241. Groups of axon are called

a) Ganglion b) Nerve sheath

c) Ganglia d) Neuropile

242. When a nerve impulse reaches the synapsis it releases a chemical messenger that diffuses across the synapsis and triggers a new impulse in the dendrites or connecting neurons. The chemical messenger is known as

a) Nerve transmitter b) Cell transmitter

c) Neurotransmitter d) All of the above

243. Acetylcholine, dopamine are examples of

a) Nerve transmitter in insect b) Cell transmitter in insect

c) Neurotransmitter in insect d) All of the above

244. Mandibles, maxillae, labium, hypopharynx, salivary glands, and neck muscles are innervated by

a) Protocerebrum b) Deutocerebrum

c) Tritocerebrum d) Sub oesophageal ganglion

245. If Sub oesophageal ganglion is removed from an insect then the insect won't be able to

a) Fly b) Eat

c) Jump d) See

246. The non-conducting cells that surrounds and protect the neurons are
   a) Ganglion b) Nerve cells
   c) Glial cells d) None of the above
247. The part of neuron which is long slender projection and is responsible for conduction of electrical impulse is
   a) Dendrites b) Axon
   c) Collaterals d) Terminal arborization
248. The brain, ganglia, major peripheral nerves and ventral nerve cords are protected by a connective tissue called
   a) Nerve sheath b) Neurolemma
   c) Medullary sheath d) Nerve fiber
249. The specialized neuroglial cells that wrap the axons and forms a thick myelin sheath is
   a) Neurolemma b) Medullary sheath
   c) Schwann cells d) Nerve fiber
250. The insect brain is also known as
   a) Suboesophageal ganglion b) Suboesopharyngeal ganglion
   c) Supra oesophageal ganglion d) Supra oesopharyngeal ganglion
251. The conspicuous paired mushroom structure like that is formed in the Neuropile of protocerebral lobe of CNS is
   a) Corpora allata b) Corpora pedunculata
   c) Corpora cardiac d) Corpora pedunculus
252. The body of nerve cell is called
   a) Axon b) Dendrites
   c) Collaterals d) Perikaryon (Neurocyte)
253. The specialized region of neuron which receives the incoming signals is
   a) Axon b) Dendrites
   c) Collaterals d) Perikaryon
254. The short protoplasmic extension arising from the end of perikaryon is
   a) Axon b) Dendrites
   c) Collaterals d) Terminal arborization
255. The side branches originating from the lateral branch of axon is
   a) Axon' b) Dendrites
   c) Collaterals d) Terminal arborization

256. The outer non-cellular nerve sheath which separates the nervous system from haemolymph is

a) Neural lamella
b) Perineurium
c) Glial cells
d) Medullary sheath

257. The brain of insect is a

a) Tri lobe structure
b) Bilobe structure
c) Mono lobe structure
d) Pentalobe structure

258. The fore brain of insect is also known as

a) Protocerebrum
b) Deutocerebrum
c) Tritocerebrum
d) Hexacerebrum

259. The mid brain of insect is also known as

a) Protocerebrum
b) Deutocerebrum
c) Tritocerebrum
d) Hexacerebrum

260. The hind brain of insect is also known as

a) Protocerebrum
b) Deutocerebrum
c) Tritocerebrum
d) Hexacerebrum

261. The stomodeum and dorsal vessel of insect are innervated by

a) Anterior sympathetic nervous system
b) Posterior sympathetic nervous system
c) Caudal sympathetic nervous system
d) Lateral sympathetic nervous system

262. Spiracles and muscles are innervated by

a) Anterior sympathetic nervous system
b) Posterior sympathetic nervous system
c) Caudal sympathetic nervous system
d) Lateral sympathetic nervous system

263. Hindgut and malpighian tubules are innervated by

a) Anterior sympathetic nervous system
b) Posterior sympathetic nervous system
c) Caudal sympathetic nervous system
d) Lateral sympathetic nervous system

264. Legs and flights muscles are innervated by
   a) Anterior sympathetic nervous system
   b) Posterior sympathetic nervous system
   c) Thorasic ganglia
   d) Lateral sympathetic nervous system

265. The synaptic gap between the axon of one neuron and collateral or terminal arborization of another neuron is
   a) 100 A° b) 100 μm
   c) 100 mm d) 100μ

266. Hypocerebral ganglion is a part of
   a) Anterior sympathetic nervous system
   b) Posterior sympathetic nervous system
   c) Caudal sympathetic nervous system
   d) Lateral sympathetic nervous system

267. Insect brain lies
   a) Above oesophagus b) Below oesophagus
   c) Behind oesophagus d) Above alimentary canal

268. The central nervous system of an insect is distinguished into
   a) 2 regions b) 3 regions
   c) 4 regions d) 5 regions

269. The sympathetic nervous system of an insect is distinguished into
   a) 2 regions b) 3 regions
   c) 4 regions d) 5 regions

270. The nerve fibres connecting the ganglia longitudinally is called
   a) Connectives b) Commissures
   c) Neurilemma d) Neurolemma

271. The nerve fibres connecting the ganglia transversely is called
   a) Connectives b) Commissures
   c) Neurilemma d) Neurolemma

272. Which one of the following controls swallowing movement in insect
   a) Frontal ganglia b) Ingluvial ganglia
   c) Recurrent nerves d) Hypocerebral ganglion

273. Which one of the following controls swallowing movement in insect

a) Frontal ganglia b) Ingluvial ganglia

c) Recurrent nerves d) Hypocerebral ganglion

274. The unstimulated state of axon is called

a) Action potential b) Reduction potential

c) Resting potential d) Depolarization

275. At resting potential the concentration of Na ions and K ions are

a) High in outside and inside respectively

b) Low in outside and inside respectively

c) High in outside and low in inside respectively

d) Low in outside and low in inside respectively

276. The nerve membrane at resting potential is

a) Negative inside and positive outside

b) Negative inside and outside

c) Positive inside and outside

d) Positive inside and negative outside

277. The resting potential of axon is

a) -70 A° b) -70μ

c) -70 mv d) -70Φ

278. At depolarization, by the action of action potential Na ion are pump inside and the nerve membrane becomes

a) Negative inside and positive outside

b) Negative inside and outside

c) Positive inside and outside

d) Positive inside and negative outside.

279. Insect nervous system is

a) Androgenic b) Chlorigenic

c) Antichlorigenic d) Cholinergic

280. Conduction of nerve impulse in insect nervous system is an

a) Electro-chemical process. b) Chemical process

c) Electrical process d) Both b and c

3

Sense Organs of Insects

281. Which one of the following are the mechanoreceptors in insect
    a) Trichoid sensilla
    b) Campaniform sensilla
    c) Chordotonal organs
    d) All of the above
282. Chordotonal organs are also known as
    a) Scoloparia
    b) Scolopophores
    c) Scolopidia
    d) Scolopale
283. Chordotonal organs are composed of specialized sensilla called
    a) Scoloparia
    b) Scolopophores
    c) Scolops
    d) Scolopale
284. The specialized sensilla in Chordotonal organ contains conspicuous sensory body called
    a) Scoloparia
    b) Scolopophores
    c) Scolopidia
    d) Scolopale
285. A highly specialized type of Chordotonal organ in the second antennal segment of an insect is
    a) Mullers organ
    b) Pearmans organ
    c) Palmans organ
    d) Johnstons organ
286. Chordotonal organ located in the proximal tibia of each leg is
    a) Subgenual organ
    b) Mullers organ
    c) Palmans organ
    d) Pearmans organ
287. A swelling like structures formed by scolopophores in the family Acridoidea which has auditory function is
    a) Mullers organ
    b) Pearmans organ
    c) Palmans organ
    d) Johnstons organ

288. The mechanoreceptor found in the cephalic tracheal system of Ephemeropteran nymphs which functions as a statocyst is

a) Mullers organ
b) Pearmans organ
c) Palmans organ
d) Johnstons organ

289. Tympanal organs are

a) Mechanoreceptors
b) Auditory organs
c) Chemoreceptors
d) Temperature receptors

290. A sound-producing structure located in the hind coxae of some males of Psocids which functions in attracting the females is

a) Mullers organ
b) Pearmans organ
c) Palmans organ
d) Johnstons organ

291. In Acridoidea (grasshoppers and locusts) family, the tympanal organ is present in the

a) 1st abdominal segment
b) 2nd abdominal segment
c) Base of fore tibia
d) Base of hind tibia

292. In Tettigonidae (crickets, mole crickets and katydids) family, the tympanal organ is present in the

a) 1st abdominal segment
b) 2nd abdominal segment
c) Base of fore tibia
d) Base of hind tibia

293. In many Lepidopterans , tympanal organs are situated at

a) 1st abdominal segment
b) Each side of metathorax
c) Base of the abdomen
d) Both b and c

294. In Neuroptera tympanal organs are located at the

a) Hind wings
b) Base of forewings
c) Forelegs
d) Hindlegs

295. The auditory organ sensitive to ultrasonic frequencies found in the head of Sphingidae family is

a) Patagia
b) Pilifer
c) Osmeterium
d) Corium

296. Johnstons organ are also know to function as auditory in insects like

a) Male mosquitoes and midges
b) Robber flies
c) Hover flies
d) Nymphalids

297. Maxillary palps of lepidopteran larva contains
   a) Sensilla basiconica b) Sensilla trichoidea
   c) Sensilla coeloconica d) Sensilla placodea

298. The olfactory receptors present in the antennae of *Apis* is
   a) Sensilla basiconica b) Sensilla trichoidea
   c) Sensilla coeloconica d) Sensilla placodea

299. Number of dorsal ocelli present in an insect
   a) 2 b) 3
   c) 4 d) 5

300. In some families of Lepidoptera, sensory organs are located on the head in the form of setose tubercles which are known as
   a) Stemmata b) Pilifers
   c) Chaetosemata d) Coremata

301. The degenerated ocelli in Blattaria is represented by
   a) Fenestra b) Fenestrae
   c) Stemmata d) All of the above

302. The dorsal ocelli is located in which part of the head
   a) Frons and Genae b) Frons
   c) Frons and vertex d) Vertex and Genae

303. Lateral ocelli or Stemmata is a characteristic feature of
   a) Exopterygota larvae b) Endopterygote larvae
   c) Exopterygota adult d) Endopterygote adult

304. In an insect eye, the outermost transparent layer which acts as a lens is
   a) Cornea b) Corneagen layer
   c) Crystalline cone d) Retina

305. The part of the epidermis extending beneath the cornea is
   a) Cornea b) Corneagen layer
   c) Crystalline cone d) Retina

306. The type of compound eyes in which each ommatidium contains a hard, refractive, conical body secreted within the crystalline cone cells is
   a) Eucone eyes b) Pseudocone eyes
   c) Acone eyes d) Exocone eyes

307. The type of compound eyes in which the cone cells are filled with transparent viscous liquid is

a) Eucone eyes
b) Pseudocone eyes
c) Acone eyes
d) Exocone eyes

308. The type of compound eyes in which the crystalline cone cells do not secrete refractive material is

a) Eucone eyes
b) Pseudocone eyes
c) Acone eyes
d) Exocone eyes

309. The type of compound eyes in which the crystalline cone is replaced by a cone-shaped extension is

a) Eucone eyes
b) Pseudocone eyes
c) Acone eyes
d) Exocone eyes

310. Primary pigment cells are present in

a) Apposition type
b) Superposition type
c) Both a and b
d) None of the above

311. Most diurnal insects have

a) Apposition type
b) Superposition type
c) Both a and b
d) None of the above

312. In superposition type primary segments are absent and image formed are indistinct, blurred and are found in

a) Diurnal insects
b) Nocturnal insects
c) Crepuscular insects
d) Both b and c

313. In lepidopteran moths, the ommatidea is covered by a tracheal layer called

a) Septum
b) Tapetum
c) Corneagen
d) Medullary sheath

314. The condition when the eyes of the two sides are contiguous is said to be

a) Holoptic
b) Dichoptic
c) Monoptic
d) Dioptic

315. The condition when the eyes of the two sides are separated is said to be

a) Holoptic
b) Dichoptic
c) Monoptic
d) Dioptic

316. Holoptic condition of compound eyes are found in

a) Zygoptera
b) Anisoptera
c) Diptera
d) Dermaptera

317. Dichoptic condition of compound eyes are found in

a) Zygoptera
b) Anisoptera
c) Diptera
d) Dermaptera

318. Semper's cells are associated with

a) Crystalline cone cells
b) Corneagen cells
c) Retinula
d) Primary cells

319. Presence of compound eyes in larval stages is a characteristic feature of

a) Mallophaga
b) Diptera
c) Coleoptera
d) Mecoptera

320. In which of the following family of insect the compound eyes are divided into a higher part that is above water level when it is floating passively and a lower part that is below water level

a) Amphizoidae
b) Haliplidae
c) Dytiscidae
d) Gyrinidae

321. The only insects known to navigate themselves by the Milky Way is

a) Dung beetles
b) Bees
c) Wasp
d) Hornets

4

Digestive System of Insects

322. The alimentary canal is longer in

a) Phytophagous insects b) Carnivorous insects

c) Both a and b d) Saprophytic insects

323. Insect foregut is also termed as

a) Stomodaeum b) Mesenteron

c) Proctodaeum d) Posterior invagination

324. Insect midgut is also termed as

a) Stomodaeum b) Mesenteron

c) Proctodaeum d) Posterior invagination

325. Insect hindgut is also termed as

a) Stomodaeum b) Mesenteron

c) Proctodaeum d) Anterior invagination

326. Which one of the following layer is the innermost

a) Epithelia layer b) Basement membrane

c) Circular muscles d) Intima

327. The portion lying between mouthparts and labrum is

a) Cibarium b) Salivarium

c) Hypopharynx d) Preoral cavity

328. Insect with mandibulate mouthparts, the hypopharynx divides the preoral cavity into

a) Ventral cibarium and ventral salivarium

b) Dorsal cibarium and dorsal salivarium

c) Ventral cibarium and dorsal salivarium

d) Dorsal cibarium and ventral salivarium

329. The region between mouth and oesophagous is
   a) Pharynx b) Crop
   c) Gizzard d) Preoral cavity
330. The organ of food reservoir in insects is
   a) Pharynx b) Crop
   c) Gizzard d) Preoral cavity
331. The gizzard is reduced to a valve in
   a) Honeybees and most dipterans b) Wasps
   c) Hornets d) Yellow jackets
332. The gizzard is modified into spine like in
   a) Aphids b) Flea
   c) Lice d) Ticks
333. The valve lying between the junction of foregut and midgut is
   a) Pyrolic valve b) Oesophageal valve
   c) Cardiac valve d) Both b and c
334. In lepidopteran larvae which of the following gland secrete silk
   a) Labial gland b) Mandibular gland
   c) Hypopharyngeal gland d) All of the above
335. Number of silk producing glands in Lepidoptera and Hymenoptera is
   a) One pair b) Two pairs
   c) Three pairs d) Four pairs
336. In lepidopteran larvae which of the following gland secrete saliva
   a) Labial gland b) Mandibular gland
   c) Hypopharyngeal gland d) All of the above
337. Number of salivary glands in insect is
   a) 1 pair b) 2 pairs
   c) 3 pairs d) 4 pairs
338. The food is taken down into the pharynx by the action of
   a) Salivarium b) Cibarial muscles
   c) Hypopharynx d) Epipharynx
339. Preoral cavity between labral epipharynx and hypo epipharynx is
   a) Salivarium b) Pharynx
   c) Cibarium d) Oesophagous

340. A median lobe present on the posterior /ventral surface of labrum or clypeus is

a) Hypopharynx b) Salivarium

c) Epipharynx d) Cibarium

341. Preoral cavity between hypopharynx and labium is

a) Salivarium b) Pharynx

c) Cibarium d) Oesophagous

342. Pharynx is modified into sucking pump in

a) Phytophagous insects b) Carnivorous insects

c) Both a and b d) Saprophytic insects

343. Which one of the following part of insect foregut acts as a food reservoir

a) Oesophagous b) Pharynx

c) Gizzard d) Crop

344. Proventriculus is also known as

a) Oesophagous b) Pharynx

c) Gizzard d) Crop

345. The conversion of nectar in honeybee takes place in

a) Oesophagous b) Pharynx

c) Gizzard d) Crop

346. 6 chitinous plates are present in the gizzard of

a) Cockroach b) Cricket

c) Grasshoppers d) All of the above

347. The food are grinded by the action of

a) Crop b) Oesophagous

c) Proventriculus d) Midgut

348. The insect of midgut is composed of how many types of epithelial cells

a) 1 b) 2

c) 3 d) 4

349. The function of gastric caeca is

a) Increase area of absorption b) Decrease area of absorption

c) Increase oxygen uptake d) Decrease oxygen uptake

350. Gastric caeca is also known by the other names such as

a) Enteric caeca
b) Hepatic caeca
c) Endoric caeca
d) Both a and b

351. Secretory cells of midgut is also known as

a) Columnar cells
b) Goblet cells
c) Calciform cells
d) Regenerative cells

352. The function of columnar cells is

a) Secrete digestive enzymes
b) Absorption
c) Transport K ions from haemolymph to lumen
d) Both a and b

353. The function of goblet cells/calciform cells is

a) Secrete enzymes
b) Absorption
c) Transport K ions from haemolymph to lumen
d) Development of epithelial cells

354. The regenerative cells of midgut which are grouped into clusters are commonly known as

a) Nidi
b) Microvilli
c) Peritrophic membrane
d) Columnar cells

355. The pH of insect midgut is usually

a) 4-5
b) 5-6
c) 6-8
d) 8-9

356. Midgut of insect is also known as

a) Proventriculus
b) Proctodaeum
c) Stomodeum
d) Ventriculus

357. Intima is absent in

a) Foregut
b) Midgut
c) Hindgut
d) All of the above

358. The midgut of insect is lined with

a) Cuticulin
b) Peritrophic membrane
c) Microvilli
d) Medullary sheath

359. Peritrophic membrane is made of clusters of cells which functions as
   a) Permeable membrane b) Semi permeable membrane
   c) Both a and b d) Selectively permeable
360. The flow of materials from the midgut into the hindgut is regulated by
   a) Stomadael valve b) Cardiac valve
   c) Oesophageal valve d) Pyrolic valve
361. Thc flow of matcrials from thc forcgut into thc midgut is rcgulatcd by
   a) Stomadael valve b) Cardiac valve
   c) Oesophageal valve d) All of the above
362. Peritrophic membrane is absent in fluid feeders except
   a) Adult *Pieris* b) Aphids
   c) Scales d) Mealybugs
363. Peritrophic membrane is absent in solid feeders in case of
   a) Ants b) Adult lepidoptera
   c) Hemiptera d) All of the above
364. The valve lying between the junction of hindgut and midgut is
   a) Pyrolic valve b) Oesophageal valve
   c) Cardiac valve d) Both b and c
365. Pyrolic region are well differentiated in
   a) Coleoptera b) Larvae of lepidoptera
   c) Diptera d) Both a and b
366. Which part of hindgut in wood eating termites and larval scarabaeidshave symbiotic function
   a) Ileum b) Colon
   c) Rectum d) Rectal pad
367. The cells of ileum in larval *Ostrinia nubilalis* undergo cyclic changes secreting a hormone called
   a) Ecdysone b) Juvenile hormone
   c) Proctodone d) All of the above
368. The hormone secreted by the larva of *Ostrinia nubilalis* has a function in
   a) Moulting b) Pupation
   c) Diapause d) Hatching

369. The absorption of salt and water from the faeces is carried out by

a) Ileum b) Colon

c) Rectum d) Rectal pads

370. The function of rectal gland found in hindgut is

a) Transport of oxygen b) Salt and water absorption

c) Secrete enzymes d) Protect from abrasion

371. The digestive enzymes in insects are secreted by

a) Salivary glands b) Columnar cells

c) Both a and b d) None of the above

372. The phenomenon whereby partial digestion occurs in the host body prior to food ingestion is known as

a) Intra intestinal digestion b) Extra intestinal digestion

c) Hypo intestinal digestion d) Hyper intestinal digestion

373. Extra intestinal digestion are common in

a) Plant bugs b) Beetles

c) Crickets d) Locusts

374. Food is moved down the gut by the contraction of muscles called

a) Peristalsis b) Porosis

c) Phagocytis d) Pinocystis

375. The function of hindgut is

a) To house symbionts b) Water absorption

c) Both a and b d) Enzyme production

376. Which part of alimentary canal is the main site of digestion

a) Foregut b) Midgut

c) Hindgut d) Oesophagous

377. The feeding of larval scarabaeiodea on rotten woods are digested by

a) Protozoa b) Bacteria

c) Fungi d) Nematode

378. The enzyme cellulase breaks down cellulose into cellobiose units and are further hydrolyzed by

a) α- glucosidase b) β- glucosidase

c) γ- glucosidase d) λ- glucosidase

379. Proteins such as keratin and collagen which are present in animals can be digested by some insects such as

a) Dermistid larvae | b) Mallophaga
c) Aphids | d) Both a and b

380. The protein keratin is found in

a) Wool | b) Hair
c) Feathers | d) All of the above

381. The digestion of protein keratin occurs by the action of

a) Invertase | b) Lipase
c) Protease | d) Keratinase

382. The complex polysaccharide cellulose are broken down in the insects by the action of

a) Collagenase | b) Keratinase
c) Proteases | d) Cellulase

383. The breakdown of carbohydrate collagen occurs by the action of

a) Collagenase | b) Keratinase
c) Proteases | d) Cellulase

384. The process of digestion involving collagenase is seen in

a) *Rhodnius* | b) *Drosophila*
c) *Lucilia* | d) *Attes*

385. Those cells which house for symbiotic microorganism are called

a) Phagocytes | b) Pinocytes
c) Mycetocytes | d) Monocytes

386. The organ formed by the aggregation of Mycetocytes is called

a) Apodeme | b) Mycetome
c) Phragma | d) Epididymis

387. In cockroach the enzyme amylase in the saliva is responsible for the digestion of

a) Starch | b) Cellulose
c) Trehalose | d) Collagen

388. Which enzyme is responsible for the breakdown of sucrose in honeybee

a) Invertase | b) Amylase
c) Cellobase | d) Pectinase

389. The stylet penetration and extra intestinal digestion is aided by the action of

a) Invertase
b) Amylase
c) Cellobase
d) Pectinase

390. Indole Acetic Acid is present in the saliva of

a) Midges
b) Aphids
c) Wasps
d) Ants

391. The type of enzyme secretion in which the cells of epithelial disintegrate during the process is termed

a) Holocrine secretion
b) Mesocrine secretion
c) Merocrine secretion
d) Endocrine secretion

392. The type of enzyme secretion in which secretion occurs without the breakdown of cells is termed

a) Holocrine secretion
b) Mesocrine secretion
c) Merocrine secretion
d) Endocrine secretion

393. Insects cannot synthesize

a) Lipids
b) Fatty acids
c) Phospholipids
d) Sterols

394. Carbohydrates are required in high amount by the

a) Xylem feeders
b) Phloem feeders
c) Scavengers
d) All of the above

395. Ectosymbiotic fungi associated with ambrosia beetle is

a) *Fusarim*
b) *Ambrosiella*
c) *Ceratocyctis*
d) Both a and b

396. Ectosymbiotic fungi associated with bark beetle is

a) *Fusarium*
b) *Ambrosiella*
c) *Ceratocyctis*
d) Both a and b

397. The breakdown of cellulose with the presence of intestinal flagellates are found in

a) Lower termites
b) Higher termites
c) Formicidae
d) Beetles

398. Higher termites belong to which family

a) Mastotermitidae
b) Kalotermitidae
c) Serritermitidae
d) Termitidae

399. Which of the following subfamily of Termitidae are known to cultivate fungal garden

a) Macrotermitinae
b) Minitermitinae
c) Termopsinae
d) All of the above

400. Endosymbiotic bacteria are found in the hindgut of

a) Cockroaches
b) Crickets
c) Scarab beetles
d) All of the abovc

401. The mycetocyte organism present in Anobiidae and cerambycidae is

a) Bacteria
b) Yeast
c) Nematode
d) Protozoa

402. Diet which are prepared artificially as that of natural forms is called

a) Holidic diet
b) Meridic diet
c) Oligic diet
d) Defined diet

403. Diet containing one or more crude form from plants, animals or microorganism is

a) Holidic diet
b) Meridic diet
c) Oligic diet
d) Defined diet

404. Which diet is mostly used in the laboratory for rearing of insects

a) Holidic diet
b) Meridic diet
c) Oligic diet
d) Defined diet

405. Diet mostly used for nutritional studies is

a) Holidic diet
b) Meridic diet
c) Oligic diet
d) Defined diet

406. Trans ovarial transmission of microbial organism is seen in

a) *Rhodnius prolixus*
b) *Stegobium paniceum*
c) *Lasioderma serricorne*
d) All of the above

5

Respiratory System of Insects

407. Respiration in insects occurs by means of internal air tubes known as
    a) Spiracles
    b) Tracheae
    c) Gills
    d) All of the above
408. The finest branches of trachea which are ramified through the organs of the body and appendages are termed
    a) Tracheoles
    b) Gills
    c) Branchiae
    d) Plastron
409. Tracheal system is absent in
    a) Most collembola
    b) Protura
    c) Some endoparasitic hymenoptera and Diptera larvae
    d) All of the above
410. Spiracles are located at which side of the region
    a) Pleura
    b) Lateral Terga
    c) Sterna
    d) Terga and sterna
411. 12 pairs of spiracles are present in the embryos of
    a) *Drosophila*
    b) *Rhodnius*
    c) *Atta*
    d) *Leptinotarsa*
412. The most primitive arrangement of functional spiracles is
    a) Holopneustic system
    b) Hemipneustic system
    c) Apneustic system
    d) Hypopneustic system
413. The type of respiration system found in Diplura is
    a) Hypopneustic
    b) Propneustic
    c) Metapneustic
    d) Hyperpneustic
414. Peripneustic form of respiratory system is found in
    a) Mostly terrestrial larvae
    b) Symphyta
    c) Bibionidae
    d) All of the above

415. Amphipneustic forms of respiratory system are found in

a) Larval Diptera
b) Pupal Diptera
c) Larval coleoptera
d) Pupal coleoptera

416. Propneustic forms of respiratory system are found in

a) Larval Diptera
b) Pupal Diptera
c) Larval coleoptera
d) Pupal coleoptera

417. Metapneustic forms of respiratory system are found in

a) Cyclorrhapha
b) Culicidae
c) Dytiscidae
d) All of the above

418. The oilgopneustic system consists of

a) Propneustic system
b) Metapneustic system
c) Amphipneustic system
d) All of the above

419. The oilgopneustic denotes the adaptive modification of life in

a) Air
b) Liquid
c) Semi-Liquid medium
d) Both b and c

420. The condition where none of the spiracles are functioning belongs to

a) Apneustic system
b) Hypopneustic system
c) Hyperpneustic system
d) Hemipneustic system

421. The condition where one or more pairs of spiracles disappeared completely belongs to

a) Apneustic system
b) Hypopneustic system
c) Hyperpneustic system
d) Hemipneustic system

422. Hypopneustic respiration system is found in

a) Mallophaga
b) Siphunculata
c) Diplura
d) Both a and b

423. Apneustic system is a characteristic feature of

a) Aquatic forms
b) Endoparasitic forms
c) Both a and b
d) Terrestrial form

424. The sclerite surrounding the spiracles which aid in resisting compressionis called

a) Atrium
b) Vestibule
c) Peritreme
d) Taenidium

425. The taenidia is absent in

a) Atrium b) Tracheoles

c) Trachea d) Both a and b

426. Which of the following intima lining is not shed during ecdysis

a) Trachea b) Spiracles

c) Tracheoles d) Both a and b

427. The thin walled vesicles formed by the dilation of tracheae in winged insects is

a) Atrium b) Vestibule

c) Alveolus d) Air sac

428. The function of air sac is

a) Reservoir of oxygen b) Maintains shape in female

c) Aid in flight d) All of the above

429. The ventilation system in insect occurs in the form of

a) Dorsal compression b) Tergal compression

c) Dorsoventral compression d) Dorsolateral compression

430. Respiratory siphon are found in

a) Metapneustic system b) Apneustic system

c) Amphipneustic system d) Propneustic system

431. Respiratory siphon are found in

a) Metapneustic system b) Apneustic system

c) Amphipneustic system d) Propneustic system

432. Insect with respiratory siphon features includes

a) Culicidae larva b) Stratiomyidae larva

c) Ptychopteridae larva d) All of the above

433. Respiratory horns are characteristic features of

a) Nematoceran pupa b) Brachyceran pupa

c) Cyclorrhaphan pupa d) Coleopteran larva

434. Plastron respiration is found in

a) Heteropteran b) Beetles

c) Both a and b d) Dipteran

435. Set of hydrofuge hairs of epicuticle holding a thin film of air is called
  a) Plastron b) Tracheal gills
  c) Blood gills d) Spiracular gills

436. Permanent physical of an insect is also known as
  a) Plastron b) Spiracular gills
  c) Blood gills d) Tracheal gills

437. The process of replacement of liquids by gases inside the tracheal system of an insect is called
  a) Diffusion b) Mass flow
  c) Osmosis d) Pneumatisation

438. Rectal gills are found in
  a) Anisoptera nymphs b) Zygoptera nymphs
  c) Ephemeroptera d) Plecoptera

439. Caudal gills are characteristic future of
  a) Anisoptera nymphs b) Zygoptera nymphs
  c) Ephemeroptera d) Plecoptera

440. Respiratory organ such as Spiracular gills are characteristic feature of
  a) Pupa of beetles b) Larva of *Myxophagan*
  c) Dipteran d) All of the above

441. Respiration by blood gills occurs in
  a) Larval Trichoptera b) Larval Chironimids
  c) Larval Tipulids d) All of the above

442. Which of the following respiratory organs are adaptive for both aerial and aquatic
  a) Tracheal gills b) Plastron
  c) Spiracular gills d) Blood gills

443. Cutaneous gas exchanges occurs in
  a) Aquatic insects b) Endoparasitic larvae
  c) Both and b d) Adult insect

444. Number of main tracheal trunks in an insect is
  a) 3 b) 4
  c) 5 d) 6

445. Which is the longest tracheal trunk and connects all the spiracles
    a) Lateral longitudinal trunk b) Dorsal longitudinal trunk
    c) Visceral longitudinal trunk d) Ventral longitudinal trunk
446. Which tracheal trunk supplies oxygen to heart
    a) Lateral longitudinal trunk b) Dorsal longitudinal trunk
    c) Visceral longitudinal trunk d) Ventral longitudinal trunk
447. Which tracheal trunk supplies oxygen to alimentary canal
    a) Lateral longitudinal trunk b) Dorsal longitudinal trunk
    c) Visceral longitudinal trunk d) Ventral longitudinal trunk
448. Which tracheal trunk supplies oxygen to nervous system
    a) Lateral longitudinal trunk b) Dorsal longitudinal trunk
    c) Visceral longitudinal trunk d) Ventral longitudinal trunk
449. The process of respiration which occurs by aeroscopic plate are found in
    a) Encyrtidae b) Braconidae
    c) Aphididae d) Ichneumonidae
450. Telescopic terminal siphon for respiration is found in
    a) Water scorpion b) Water boatman
    c) Water strider *d) Eristalis* larva
451. Caudal breathing tube is a characteristic feature of
    a) Water scorpion b) Water boatman
    c) Water strider d) *Eristalis*
452. Respiration in insect occurs by
    a) Chemical phases b) Physical phases
    c) Biological phases d) Both a and b
453. The web likc cell present at the base of group of Tracheoles is
    a) End cell b) Basement cell
    c) Platform cell d) Semper's cell
454. Changes in the volume of tracheal system is called
    a) Ventilation b) Pneumatisation
    c) Expiration d) Diffusion

6

Circulatory System of Insects

455. The principle pulsatory organ in insect is
   a) Heart b) Aorta
   c) Accessory pulsatory organ d) All of the above
456. Sac-like structure situated in various region of the body which function in adequate circulation of blood is
   a) Heart b) Aorta
   c) Accessory pulsatory organ d) All of the above
457. The circulatory system of insect is
   a) Closed type b) Open type
   c) Absent d) Both an and b
458. Most of the circulation in insect occurs in
   a) Cavities of the body/ haemocoele
   b) Arteries
   c) Aorta
   d) Heart
459. The diaphragm which extends above the alimentary canal and the blood space is
   a) Dorsal diaphragm b) Ventral diaphragm
   c) Dorsoventral diaphragm d) Dorsolateral diaphragm
460. The diaphragm which extends above the ventral nerve cord and stretches across abdominal cavity is
   a) Dorsal diaphragm b) Ventral diaphragm
   c) Dorsoventral diaphragm d) Dorsolateral diaphragm
461. The ventral diaphragm is limited by
   a) Pericardial sinus b) Perineural sinus
   c) Perivisceral sinus d) Visceral sinus

462. The dorsal diaphragm is limited by
   a) Pericardial sinus
   b) Perineural sinus
   c) Perivisceral sinus
   d) Visceral sinus

463. The central cavity or sinus between dorsal and ventral sinus which contain the principal internal organs is
   a) Pericardial sinus
   b) Perineural sinus
   c) Perivisceral sinus
   d) Alary muscles

464. The dorsal vessel of circulatory system in insect extends from
   a) Caudal end of the body to thorax and terminates in the head
   b) Thorax to head
   c) Abdomen to head
   d) Abdomen to heart

465. The regions that constitute the dorsal vessels are
   a) Heart and aorta
   b) Aorta only
   c) Heart and nerve cord
   d) All of the above

466. The pumping organ and conducting vessel of insect circulatory system are
   a) Heart and nerve cord respectively
   b) Heart and abdomen respectively
   c) Heart and aorta respectively
   d) Heart and spiracles respectively

467. The position of the heart of insect lie within
   a) Perineural sinus
   b) Pericardial sinus
   c) Perivisceral sinus
   d) Visceral sinus

468. The termination of aorta is at
   a) Protocerebrum
   b) Deutocerebrum
   c) Tritocerebrum
   d) Suboesophageal ganglion

469. Histologically insect heart is composed of
   a) Single layer of striated muscle
   b) Double layer of striated muscle
   c) Triple layer of striated muscle
   d) Single unstriated muscle

470. The blood of insect contains
   a) Plasma
   b) Haemocytes
   c) Both a and b
   d) None of the above

471. Blood enters the insect heart through

a) Lateral inlets/incurrent ostia b) Auricular valve

c) Excurrent ostia d) Alary muscle

472. Vertical slit like opening located in the lateral wall of the heart is

a) Lateral inlets/incurrent ostia b) Auricular valve

c) Excurrent ostia d) Alary muscle

473. Entry of blood into the heart takes place at

a) Systole b) Contraction

c) Diastasis d) Diastole

474. The flow of haemolymph outward is prevented at which stage

a) Systole b) Relaxation

c) Diastasis d) Diastole

475. Relaxation of heart occurs at which stage

a) Systole b) Diastasis

c) Diastole d) Both a and b

476. The entry of blood is prevented and haemolymph is force out by the presence of

a) Auricular valve b) Alary muscle

c) Excurrent ostia d) Incurrent ostia

477. The anterior prolongation of the dorsal vessel which functions as the principle artery is

a) Alary muscle b) Aorta

c) Aliform muscle d) Striated muscle

478. Number of alary muscles present in the circulatory system of an insect

a) 1 pair b) 2 pairs

c) 3 pairs d) 4 pairs

479. The junction between the heart and aorta is

a) Pyrolic valve b) Auricular valve

c) Aortic valve d) Ventricular valve

480. The plasma contains

a) 45% water b) 55% water

c) 75% water d) 85% water

481. High levels of fee amino acids are found in the plasma of
   a) Lepidoptera and Hymenoptera b) Coleoptera and Diptera
   c) Diptera and Hemiptera d) Diptera and lepidoptera
482. The sugar in the blood of insect is
   a) α-Trehalose b) β-Trehalose
   c) γ-Trehalose d) Glucose
483. The major blood sugar in *Apis* is
   a) α-Trehalose b) β-Trehalose
   c) γ-Trehalose d) Glucose
484. The phenomenon of emission of blood through pore or slits when the insects are disturb is known as
   a) Reflex bleeding b) Voluntary bleeding
   c) Involuntary bleeding d) Sudden bleeding
485. The blood cells of insects is also known as
   a) Plasma b) Haemocytes
   c) Cyctocytes d) Oenocytes
486. The main function of haemocyte is
   a) Phagocytosis b) Pinocytosis
   c) Ingestion of small solid particles d) Both a and c
487. The rhythm of insect heart is purely
   a) Androgenic b) Neurogenic
   c) Myogenic d) Chlorogenic
488. The flow of haemolymph into the dorsal vessel takes place by openings called
   a) Ostia b) Aorta
   c) Septum d) Valves
489. The circulation of the haemolymph takes place
   a) Dorsoventrally b) Posteroventrally
   c) Dorsolaterally d) Dorsally
490. Which of the following function of haemolymph is false
   a) Eclosion b) Ecdysis
   c) Reflex bleeding d) Respiration

491. Which type of haemocyte plays an important role in coagulation

a) Prohaemocyte b) Cytocyte

c) Podocyte d) Plasmatocyte

492. Coagulocyte is also termed as

a) Prohaemocyte b) Cytocyte

c) Podocyte d) Plamatocyte

493. Which type of haemocyte gives rise to new type of haemocyte

a) Prohaemocyte b) Cytocyte

c) Podocyte d) Plamatocyte

494. Which type of haemocyte is also known as stem cells

a) Prohaemocyte b) Cytocyte

c) Podocyte d) Plamatocyte

495. The type of haemocyte with basophilic cytoplasm is

a) Prohaemocyte b) Cytocyte

c) Podocyte d) Plasmatocyte

496. The type of haemocyte with acidophilic cytoplasm is

a) Granulocytes b) Cytocyte

c) Podocyte d) Plamatocyte

497. The largest blood cells is

a) Oenocytes b) Cytocyte

c) Podocyte d) Plamatocyte

498. The phase in which the heart rest at expanded condition is

a) Systole b) Diastole

c) Diastasis d) Contraction

499. Adipoleucocytes type of haemocytes are found in

a) Hemipterans bug b) Dipteran

c) Coleopteran d) Lepidopteran

7

Excretory System of Insects

500. The main excretory organ of insects is
   a) Malpighian tubules
   b) Nephrocytes
   c) Fat body
   d) Labial gland
501. Malpighian tubules originates from
   a) Thorax
   b) Dorsal region
   c) Foregut
   d) Midgut and hindgut
502. Malpighian tubules was discovered by an anatomist Marcello Malpighi native of
   a) Austria
   b) Italy
   c) Denmark
   d) Sweden
503. The principal gas expelled in excretion process is
   a) Oxygen
   b) Carbon monoxide
   c) Carbon dioxide
   d) Both b and c
504. A condition in which the malpighian tubules are freely arranged distally but in some insects they are in close contact with the hindgut is known as
   a) Nephridial condition
   b) Cystonephredial condition
   c) Cryptonephredial condition'
   d) Filter chamber
505. The function of Cryptonephredial in insect is
   a) Transport of oxygen
   b) Transport of nutrient
   c) Water conservation by withdrawing it from the faeces
   d) Transport of ions
506. Cryptonephredial condition is found in
   a) Larvae and adults of Coleoptera
   b) Larva of Lepidoptera
   c) Larvae of Diptera
   d) Both a and b

507. Malpighian tubules are represented by the presence of papillae in

a) Certain Diptera b) Protura

c) Strepsitera d) All of the above

508. Malpighian tubules are absent in insects such as

a) Collembola b) Aphids

c) Diplura d) All of the above

509. The shape of malpighian tubule as arborescent like is found in

a) *Galleria mellonella* b) *Melolontha*

c) *Japyx* d) *Donacia*

510. Closely packed diverticula type of malpighian tubules occurs in

a) *Galleria mellonella* b) *Melolontha*

c) *Donacia* d) *Japyx*

511. The type of tubules which converge in to a common ureter is found in

a) *Galleria mellonella* b) *Melolontha*

c) *Donacia* d) *Acheta* (Gryllidae)

512. Tubules which are arranged into a common ampulla occurs in

a) *Haltica* b) *Donacia*

c) *Rhodnius* d) Both a and b

513. Malpighian tubules are composed of

a) 1-4 epithelial cells b) 2-4 epithelial cells

c) 2-5 epithelial cells d) 3-8 epithelial cells

514. The main excretory product in insect is

a) Urea b) Allantoin

c) Uric acid d) Ammonia

515. The fat bodies of an insect are irregular mass of cells known as

a) Nephrocytes b) Tropocytes

c) Trophocytes d) Oenocytes

516. The cells of fat body of which insect larva is composed of liver organelles containing catalase and uric oxidase

a) *Sialis* b) *Calliphora*

c) *Lucilia* d) *Calpodes*

517. The deposition of uric acid and urates in the urate cells of fat body occurs in

a) Collembola b) *Blatta*

c) Parasitic hymenoptera d) All of the above

518. The process of excretion by Labial gland as the excretory organ occurs in

a) Collembola b) Protura

c) Thysanura d) All of the above

519. The function of fat body cells is

a) Transport of nutrients b) Protection

c) Homeostasis d) Storage and excretion.

520. Which of the following organ does not play any role in excretion process

a) Integument b) Trachea

c) Oenocytes d) Midgut

521. The type of excretion of terrestrial insect is

a) Aminotelism b) Uricotelism

c) Ureotelism d) Ammonotelism

522. The type of excretion of aquatic insects is

a) Aminotelism b) Uricotelism

c) Ureotelism d) Ammonotelism

523. The main excretory product of aquatic insect is

a) Urea b) Allantoin

c) Uric acid d) Ammonia

524. The phenomenon of light production by the presence of malpighian tubules occurs in

a) *Danaus Plexippus* b) *Laminitis archippus*

c) *Hypolimnas bolina* d) *Bolitophila*

525. The production of silk by the neuropteran larva is secreted by

a) Labial gland b) Mandibular gland

c) Salivary gland d) Malpighian tubules

526. The cells which removes foreign materials of higher molecular weight that cannot be removed by the malpighian tubules is

a) Tropocytes b) Oenocyte

c) Nephrocytes d) Binucleate cells

527. The ion actively involved in excretion is

a) K ions
b) Na ions
c) Cl ions
d) Fe ions

528. The process of excretion in insect is aided by enzyme

a) Proteases
b) Lipases
c) Carbonic anhydrase
d) Invertase

529. The excretory product in the form of urea occurs in

a) Cloth moth
b) Mosquitoes
c) Red cotton bug
d) Both a and b

530. The excretory product of red cotton bug (*Dysdercus*) is

a) Urea
b) Ammonia
c) Uric acid
d) Allantoin

531. The emission of white colour from the wings of Pierid butterflies is due to the presence of

a) Urea
b) Ammonia
c) Uric acid
d) Allantoin

532. The end products of excretion are excreted out by the exuviae at each moulting, such type of excretion takes place through

a) Fat bodies
b) Labial glands
c) Wall of alimentary canal
d) Integument

533. The cells of malpighian tubules bears two region namely

a) Brush border in the proximal region and honey comb border in the distal region
b) Honey comb border in the proximal region and Brush border in the distal region
c) Convex border in the proximal region and Concave border in the distal region
d) Concave border in the proximal region and Convex border in the distal region

534. The excretory product in solid form is

a) Urea/uric acid
b) Allantoin
c) Honeydew
d) Carbon dioxide

535. The excretory product in the form of semi-solid is

a) Urea b) Allantoin

c) Uric acid d) Honey dew

536. The excretory product in gas form is

a) Urea b) Allantoin

c) Uric acid d) Carbon dioxide

537. The excretory product of blowfly insect (*Calliphora*) is

a) Allantoin b) Urea

c) Uric acid d) Ammonia

538. The breakdown of uric acid into allantoin occurs by the action of

a) Urease b) Protease

c) Lipases d) Uricase

8

Secretory System of Insects

539. Lac glands are secreted from the family
    a) Lacciferidae
    b) Coccidae
    c) Ahipdoidae
    d) Tingidae
540. Presence of structure like Pore on frontal region of head of termitethat exude defensive secretions is
    a) Phragma
    b) Fontanelle
    c) Osmeteria
    d) Coremata
541. Maxillary glands are prominent in which type of insect
    a) Collembola
    b) Protura
    c) Neuropteran larvae
    d) All of the above
542. The enzyme release by the filamentous tubes formed by the labial gland in Saturiniidae and Bombycidae is
    a) Proteases
    b) Peptidase
    c) Chitinases
    d) Cocoonase
543. Secretion of silk from the fore tarsi occurs in
    a) Embioptera
    b) Male *Hilara*
    c) Both a and b
    d) *Hydrophilus*
544. The secretion of mixture by the pygidial glands in Adephaga suborder is
    a) Mixture of Quinone and Hydrogen peroxide
    b) Mixture of Hydrogen derivatives
    c) Mixture of terpenes and quinones
    d) Mixture of cardelodes and HCN
545. The enlarged frontal gland found in insect order Isoptera are known to secrete sticky defensive substance and they are found in which caste of the colony
    a) Queen
    b) King
    c) Workers
    d) Nasute Soldiers

546. The composition of the sticky secretion from the insect order Isoptera is
   a) Terpenes and Quinones b) Terpenes and HCN
   c) Terpenes and acetic acid d) Terpenes and resins

547. The eversible repugnatorial gland found in Lepidopterean larvae is
   a) Meconium b) Ptilinum
   c) Androconia d) Osmeteria

548. Osmeteria are more pronounced in which part of the body region
   a) Thorax b) Dorsal region
   c) Abdomen (6th and 7th) d) All of the above

549. The type of gland which discharge their secretion through specialized duct or aperture is
   a) Exocrine b) Endocrine
   c) Holocrine d) Merocrine

550. The type of gland without any specialized aperture or duct and secretes their product diffusing with the blood and spreads all over the body is
   a) Exocrine b) Endocrine
   c) Holocrine d) Merocrine

551. Wax gland is a characteristic feature of which insect order
   a) Homoptera b) Heteroptera
   c) Adephaga d) Polyphaga

552. Chinese white wax are secreted by
   a) *Eriosoma lanigerum* b) *Laccifer lacca*
   c) *Ula macroptera* d) *Ericerus pela*

553. The scent glands or Androconia in Lepidoptera are characteristically
   a) Specialized scales b) Modified bristles
   c) Setae d) Spurs

554. The eversible abdominal hair pencil which are known to secrete Aphrodisiac secretion are characteristically
   a) Modified glandular scales b) Specialized scales
   c) Modified bristles d) Spurs

555. Aphrodisiac secretion are prominent in
   a) *Danaus gilippus* (Queen Butterfly)
   b) *Bombyx mori*
   c) *Portheria dispar*
   d) *Ephestia* sp.

556. Which of the following insect are known for their aromatic secretion
   a) Male blattids
   b) Female coccoidea
   c) Symphiline coleoptera
   d) *Lebia scapularis*
557. The eversible spinose tubercles located at the abdominal end occurs in the larvae of
   a) Lycaenid
   b) Braconid
   c) Ichneumonid
   d) Stylopid
558. Honey bee venom contains
   a) Cardeloides
   b) Cantharidin
   c) HCN
   d) Melithin
559. Neurosecretory cells are typically arranged in the
   a) Protocerebrum
   b) Deutocerebrum
   c) Tritocerebrum
   d) Frontal ganglion
560. The brain hormone is released by
   a) Neurosecretory cells
   b) Bursicon
   c) Juvenile hormone
   d) Neotanin
561. The process of moulting and termination of diapause in insect is due to the secretion of
   a) Ecdysone
   b) Bursicon
   c) Juvenile hormone
   d) Neotanin
562. The prothoracic gland secretes
   a) JH
   b) Ecdysone
   c) Diuretic hormone
   d) Bursicon
563. The function of corpora cardiaca includes
   a) Storage of hormones
   b) Release of hormones synthesized by Neurosecretory cells
   c) Increase in yolk deposition in the oocytes
   d) Both a and b
564. The process of hardening and darkening of cuticle is due to the secretion of hormone
   a) Ecdysone
   b) Bursicon
   c) Juvenile hormone
   d) Neotanin

565. The hormone released by Neurosecretory cells which functions in excretion and water balance is

a) Bursicon b) Juvenile hormone
c) Neotanin d) Diuretic hormone

566. Ecdysone hormone are

a) Steroid in nature b) Terpenoid in nature
c) Esters d) All of the above

567. The process of Moulting continues throughout the life in which of the following

a) Thysanura b) Coleoptera
c) Hemiptera d) Diptera

568. Enzymes rich in moulting fluid are

a) Proteases and Invertase b) Protease and Lipases
c) Proteases and Chitinases d) Chitinases and Lipases

569. Juvenile hormones are released by

a) Corpora cardiaca b) Corpora allata
c) Prothorasic gland d) Neurosecretory cells

570. Juvenile hormones are

a) Steroid in nature b) Terpenoid in nature
c) Polypeptide proteins d) All of the above

571. Metamorphosis in insect occurs due to low titre of

a) Ecdysone b) Juvenile hormone
c) Diuretic hormone d) Bursicon

572. The removal of corpora allata in adult female leads to

a) Prevention of yolk deposition b) Increase in ecdysone
c) Prevention of moulting d) All of the above

573. The Neurosecretory cells of the brain synthesis

a) Brain hormone and Bursicon
b) Prothoracicotrophic hormone and Gonadotrophic Hormone
c) Diuretic hormone
d) All of the above

574. Juvenile hormone is also known as

a) Neotanin b) Eclosion hormone

c) Adepokinetic hormone d) Hyperglycemic hormone

575. The gland formed by the fusion of corpora allata, corpora cardiaca and prothoracic gland in Cyclorrhaphan insect is

a) Palmers organ b) Verson's gland

c) Dufour's gland d) Weismann's ring

576. The precursor of Juvenile hormone in mammals is

a) Amino acid b) Proteins

c) Carbohydrates d) Cholesterol

577. The precursor of Juvenile hormone in insect is

a) Amino acid b) Proteins

c) Farnosil pyrophosphate d) Cholesterol

578. Which of the following type of juvenile hormone is present only in the eggs of *Manducasexta*

a) JH I b) JH II

c) JH III d) JH 0

579. The JH present in all Lepidopteran is

a) JH 0 and JH 1 b) JH I and JH II

c) JH II and JH III d) JH II and JH 0

580. Which type of JH helps in the synthesis of proteins

a) JH I b) JH II

c) JH III d) JH 0

581. The precursor of ecdysone in insect is

a) Amino acid b) Carbohydrates

c) Cholesterol d) Protein

582. JH functions as morphogenic hormone in which stage of the insect

a) Egg b) Imaginal

c) Pupal d) Adult

583. The social castes in honey bees occurs due to

a) JH b) Ecdysone

c) Bursicon d) Diuretic hormone

584. The phenomenon whereby the larval stage of an insect are known to exhibit adult character is known as

a) Thyletoky
b) Amphitoky
c) Prothethely
d) Arrhenotoky

585. The hormone which initiate skeletal and visceral excitation is

a) Antidiuretic hormone
b) Adepokinetic hormone
c) Gonadotrophic hormone
d) Proctolin

586. Prothoracic glands are located in which part of the region

a) Dorsolateral
b) Ventrolateral
c) Dorsally
d) Ventrally

587. The secretion of ecdysone by the prothoracic gland is activated by

a) Neurosecretory cells
b) Frontal ganglion
c) Recurrent nerves
d) Suboesophageal ganglion

588. Which of the following gland becomes deactivated at adult stage of the insect

a) Prothoracic gland
b) Corpora cardiaca
c) Corpora allata
d) Neurosecretory cells

589. The moths are distasteful due to presence of hydrogen cyanide (HCN) throughout all stages of their life cycle in

a) Zygaenidae
b) Epipyropidae
c) Dalceridae
d) Zygaenoidea

590. In Lepidopteran insects, a pair of accessory glands opens into the anterior part of silk gland which is termed as

a) Lyonnet's glands
b) Filippi's gland
c) Verson's gland
d) Both a and b

591. Fontanelles are present in which caste of termite

a) Soldiers
b) Workers
c) Drones
d) Queen

9

Reproductive System of Insects

592. In males the reproductive structures usually lies behind

a) 6th abdominal sternum
b) 7th abdominal sternum
c) 8th abdominal sternum
d) 9th abdominal sternum

593. In females the reproductive structures lies behind

a) 6th abdominal sternum
b) 7th or 8th abdominal sternum
c) 8th or 9th abdominal sternum
d) 9th or 10th abdominal sternum

594. Which of the following organ is not a part of male reproductive structure

a) Mesadenia
b) Ectadenia
c) Vasa deferentia
d) Spermatheca

595. The body form of an insect which shows both male and female characteristic features is called

a) Gynandromorphy
b) Sex mosaics
c) Polyandromorph
d) Both a and b

596. In which of the following insect order the follicle is connected with the vas deferens by a slender tube called vas efferens

a) Mallophaga
b) Siphunculata
c) Coleoptera
d) All of the above

597. The part of ,ale testicular follicle containing the primordial germ cells is

a) Germanium
b) Zone of growth
c) Zone of division
d) Zone of transformation

598. The division of follicle in male whereby the spermatogonia develops into spermatocytes is

a) Germanium
b) Zone of growth
c) Zone of division
d) Zone of transformation

599. The zone of division involves the development of
   a) Spermatogonia to spermatocytes
   b) Spermatogonia to spermatids
   c) Spermatocytes to spermatids
   d) Spermatids to spermatogonia

600. Spermatozoa are developed in which zone of male follicles
   a) Zone of reduction
   b) Zone of division
   c) Zone of growth
   d) Zone of transformation

601. Presence of large elements in the male testis are called
   a) Apical cells
   b) Verson's cells
   c) Germinal cells
   d) Both a and b

602. In some insects the vas deferens are enlarged forming a sac like structure in which spermatozoa are found in clustered arrangement. The sac like structure is termed
   a) Ejaculatory duct
   b) Aedagus
   c) Vesicular seminalis
   d) Vas deferentia

603. Which part of the male reproductive organ receives sperm from the testes and transport to ejaculatory duct
   a) Vas deferen
   b) Aedagus
   c) Seminal vesicle
   d) Vesicular seminalis

604. Which of the following is not a part of female reproductive organ
   a) Seminal vesicle
   b) Ovarioles
   c) Germarium
   d) Spermatheca

605. The finger like evagination formed at the terminal section of ejaculatory duct is
   a) Vas deferen
   b) Aedagus
   c) Seminal vesicle
   d) Vesicular seminalis

606. Male accessory glands formed as an evagination of vas deferens is
   a) Mesadenia
   b) Ectadenia
   c) Vas deferentia
   d) Seminal vesicle

607. Male accessory glands formed as an evagination of ejaculatory duct is
   a) Mesadenia
   b) Ectadenia
   c) Vas deferentia
   d) Seminal vesicle

608. Mesadenia is derived from
   a) Endoderm b) Ectoderm
   c) Mesoderm d) Both a and b

609. Ectadenia is derived from
   a) Endoderm b) Ectoderm
   c) Mesoderm d) Both a and b

610. Conglobate gland which functions in the formation of spermatophore occurs in
   a) Cockroaches b) Termites
   c) Ants d) Wasps

611. Mushroom shaped glands occurs in
   a) Male cockroach b) Male mantid
   c) Female mantid d) Both a and b

612. The inner layer of ovarioles consisting of epithelium cells are covered by
   a) Connective tissue b) Tunica propria
   c) Muscle fibres d) All of the above

613. The apex part of female ovarioles is
   a) Terminal filament b) Germarium
   c) Vitellarium d) Primordial germ cells

614. The major portion of ovarioles containing the developing eggs and nutritive cells is
   a) Terminal filament b) Germarium
   c) Vitellarium d) Primordial germ cells

615. The nurse cells or nutritive cells is also termed as
   a) Oenocytes b) Trophocytes
   c) Granulocytes d) Adipocytes

616. The type of ovarioles where nutritive cells are absent is
   a) Panoistic type b) Polytrophic type
   c) Acrotrophic type d) Meroistic type

617. The type of ovarioles whereby the nurse cells and oocytes are arranged in alternate manner is
   a) Panoistic type b) Polytrophic type
   c) Acrotrophic type d) Meroistic type

618. The type of ovarioles where the nurse cells are situated at the apices part is

a) Panoistic type
b) Polytrophic type
c) Acrotrophic type
d) Meroistic type

619. In which type of ovarioles, the oocytes can synthesize RNA, proteins and glycogen

a) Panoistic type
b) Polytrophic type
c) Acrotrophic type
d) Meroistic type

620. The process of discharging the matured eggs from the ovarioles is known as

a) Ovum
b) Ova
c) Ovulation
d) Oogenesis

621. The transfer of sperm in Ribaga's organ are found in which of the following insect

a) Cimicidae
b) Xylocoris
c) Braconids
d) Both a and b

622. Haemocoelic migration of sperms occurs in

a) *Hesperoctenes*
b) *Lyctocoris*
c) *Prostemma*
d) All of the above

623. The type of reproduction where the eggs contain enough yolk to nourish the embryos which are deposited by the mother soon after hatching is termed

a) Ovoviviparity
b) Adenotrophic viviparity
c) Haemocoelous viviparity
d) Pseudoplacental viviparity

624. The type of reproduction in which the developing embryo feeds on the secretion of female hypotrophied uterine glands and molts twice after which are deposited as a matured larva and undergoes pupation is

a) Ovoviviparity
b) Adenotrophic viviparity
c) Haemocoelous viviparity
d) Pseudoplacental viviparity

625. The type of reproduction in which the embryo develops in the maternal vagina with yolkless egg and is nourished through placenta like structure is

a) Ovoviviparity
b) Adenotrophic viviparity
c) Haemocoelous viviparity
d) Pseudoplacental viviparity

626. The type of reproduction in which the embryo develops in the haemocoele of mother is

a) Ovoviviparity
b) Adenotrophic viviparity
c) Haemocoelous viviparity
d) Pseudoplacental viviparity

627. The laying of eggs containing embryos in an advanced state is

a) Ovoviviparity
b) Adenotrophic viviparity
c) Haemocoelous viviparity
d) Pseudoplacental viviparity

628. The nourishment of embryos occurs in which type of reproduction

a) Ovoviviparity
b) Adenotrophic viviparity
c) Haemocoelous viviparity
d) Pseudoplacental viviparity

629. Facultative thelytoky is seen in

a) *Coccus hesperidium*
b) *Coccus viridae*
c) *Alloeorhynchus*
d) *Xyleborus*

630. Obligate thelytoky is seen in

a) Curculionidae
b) Psychids
c) Phasmids
d) All of the above

631. Gynogenesis or pseudogamy occurs in

a) *Luffia lapidella*
b) *Ptinus clavipes*
c) *Bacillus*
d) Both a and b

632. Functional hermaphroditism is seen in

a) *Icerya purchasi*
b) *Icerya bimaculata*
c) *Termitostroma*
d) All of the above

633. Non-functional hermaphroditism is seen in

a) Stonefly (*Perla marginata*)
b) Termites (*Neotermes zuluensis*)
c) Ants
d) Both a and b

634. The deposition of yolk in the oocyte is called

a) Vittellogenesis
b) Ovulation
c) Oogenesis
d) Oogamy

635. The development of ova is known as

a) Vittellogenesis
b) Ovulation
c) Oogenesis
d) Oogamy

636. The phenomenon of receiving and storing the sperms in female occurs in

a) Seminal vesicle
b) Spermatheca
c) Collateral gland
d) Accessory gland

637. The phenomenon of maturation of sperm within the female is known as

a) Vittellogenesis
b) Ovulation
c) Oogenesis
d) Capacitation

638. The organ of Ribaga which is present in bedbugs is located in the

a) 1st abdominal segment
b) 2nd abdominal segment
c) 3rd abdominal segment
d) 4rd abdominal segment

639. The arrangement of vas deferens into a long coiled tube which is known as epididymis and functions in storing sperms occurs in

a) Lepidoptera
b) Hymenoptera
c) Dermaptera
d) Coleoptera

640. Which hormone is required for the process of Vittellogenesis

a) Ecdysone
b) JH
c) Bursicon
d) Diuretic hormone

641. The process of producing young ones without fertilization is

a) Paedogenesis
b) Parthenogenesis
c) Polyembryony
d) Hermaphroditism

642. Cyclic Parthenogenesis occurs in

a) Bees
b) Wasps
c) Aphids
d) Ants

643. Obligate parthenogenesis occurs in

a) Bees
b) Wasps
c) Aphids
d) Ants

644. The type of parthenogenesis where only males are produced is

a) Thelytoky
b) Amphitoky
c) Arrhenotoky
d) Deuterotoky

645. The type of parthenogenesis where only females are produced is

a) Thelytoky
b) Amphitoky
c) Arrhenotoky
d) Deuterotoky

646. The type of parthenogenesis where only females are produced is

a) Deuterotoky b) Amphitoky
c) Arrhenotoky d) Both a and b

647. The process of producing more than one embryo is

a) Paedogenesis b) Parthenogenesis
c) Polyembryony d) Hermaphroditism

648. Reproduction by young immature ones is

a) Paedogenesis b) Parthenogenesis
c) Polyembryony d) Hermaphroditism

649. Larval Paedogenesis is seen in

a) Gall midge b) *Miaster*
c) *Platygaster* d) *Ptinus*

650. Larval Paedogenesis is seen in

a) Gall midge b) *Miaster*
c) *Platygaster* d) *Ptinus*

651. Polyembryony occurs in

a) Gall midge b) *Miaster*
c) *Platygaster* d) *Ptinus*

652. Which of the following reproductive part is not present in male

a) Seminal vesicle b) Accessory gland
c) Testes d) Spermatheca

653. The functional unit of ovary is

a) Ova b) Ovum
c) Ovarioles d) Germarium

654. The function of Spermatheca is to

a) Store ova b) Store spermatozoa
c) Store nutrients d) Store oxygen

655. The type of sperm transfer where the spermatophore are ejected out in to the open space by the male and then the female walks over it occurs in

a) Diplura b) Collembola
c) Thysanura d) Protura

656. Pseudocerci is also known as

a) Urogomphi
b) Cerci
c) Papilla
d) Ovipositor

657. Sac-like organ in which the spermatophore is stored immediately after mating and functions in secreting enzymes that breaks the spermatophore down into nutrients that can be used by the female is

a) Spermatheca
b) Bursa copulatrix
c) Seminal vesicle
d) Vas deferens

658. The insect group belonging to Lepidopteran order in which the females have a single genital opening for both mating and egg laying are

a) Aglossata
b) Ditrysia
c) Monotrysia
d) Zeugloptera

659. In a bee colony the females are developed from

a) Unfertilized egg
b) Fertilized egg
c) Parthenogenitically
d) Both a and b

660. In a bee colony the males are developed from

a) Unfertilized egg
b) Fertilized egg
c) Parthenogenitically
d) Both a and b

661. Traumatic insemination, also known as hypodermic insemination occurs in

a) Bed bugs
b) *Drosophila*
c) Strepsipteran
d) All of the above

10

Embryology of Insects

662. The eggs of most insects are packed with yolkin the centre of the cytoplasm, this situation is called

a) Isolecithal
b) Centolecithal
c) Meroblastic
d) Holoblastic

663. the individual cells surrounding the daughter nuclei formed after fusion of egg and sperm is called

a) Energids
b) Cleavage cells
c) Both a and b
d) Tropocytes

664. The layer formed by mitotic division of energids after its migration towards the periphery of the egg is termed

a) Blastoderm
b) Gastrulation
c) Blastokinesis
d) Germband

665. Some energids refrain from migration towards the periphery of the eggs and remain in the yolk, such energids are termed

a) Yolk cells
b) Vitellophages
c) Germarium
d) Both a and b

666. Secondary Vitellophages occurs in

a) Coleoptera
b) Diptera
c) Dermaptera
d) Orthoptera

667. Presence of lid-like structure called an operculum at the anterior pole of the egg, from which the nymph emerges during hatching occurs in

a) Embioptera
b) Dictyoptera
c) Mallophaga
d) Phasmida

668. The thickening of Blastoderm in the ventral region of the egg gives rise to

a) Blastoderm
b) Gastrulation
c) Blastokinesis
d) Germband

669. The dorsal thin part of Blastoderm which remains as an extra embryonic membrane is termed

a) Serosa b) Chorion

c) Amnion d) Energids

670. The membrane formed by the folding of periphery after formation of germ band is called

a) Serosa b) Chorion

c) Amnion d) Energids

671. The formation of germinal layers during embryonic stage is called

a) Blastoderm b) Gastrulation

c) Blastokinesis d) Germband

672. Ootheca are produce by

a) Cockroaches b) Mantids

c) Cassidinidae d) All of the above

673. The phenomenon involving the movement, rotation, revolution of the embryo within the yolk is termed

a) Blastoderm b) Gastrulation

c) Blastokinesis d) Germband

674. The types of embryonic movement includes

a) Anatrepsis only b) Katatrepsis only

c) Both a and b d) None of the above

675. The movement of embryo from the posterior to the anterior pole of egg is

a) Anatrepsis b) Katatrepsis

c) Gastrulation d) Blastokinesis

676. The movement of embryo from the ventral to the dorsal surface of the egg is

a) Anatrepsis b) Katatrepsis

c) Gastrulation d) Blastokinesis

677. The shell of an insect egg is called

a) Amnion b) Serosa

c) Chorion d) Germband

678. Eclosion is the process of

a) Rupturing of egg shell b) Hatching

c) Aestivation d) Both a and b

679. The process of rupturing of egg shell occurs through
   a) Intrinsic mechanism
   b) Extrinsic mechanism
   c) Ptilinial mechanism
   d) All of the above
680. Rupturing of egg shell through Intrinsic mechanism is seen in
   a) *Gryllus*
   b) Ants
   c) Bees
   d) Wasps
681. The eggs are called nits in insect order
   a) Anoplura (Siphunculata)
   b) Mallophaga
   c) Diptera
   d) Both a and b
682. The phenomenon of parental care in insect occurs in
   a) Belostomatidae
   b) Coreidae
   c) Reduviidae
   d) All of the above
683. Rupturing of egg shell through Ptilinial mechanism is seen in
   a) Cyclorrhaphan
   b) Nematoceran
   c) Brachyceran
   d) Chironimids
684. Rupturing of egg shell through Intrinsic mechanism is seen in
   a) Coleoptera
   b) Isoptera
   c) Diptera
   d) Lepidoptera
685. The successive form attained after emerging from the egg and in between two molts is called
   a) Imago
   b) Sub-imago
   c) Pronymph
   d) Instar
686. The stage at which the instar lie within the old unshed cuticle or an insect that has completed the metamorphosis from larvae to adult but is still within the pupa is referred to as
   a) Imago
   b) Sub-imago
   c) Pronymph
   d) Pharate
687. The change in form from embryo to adult is termed
   a) Ecdysis
   b) Eclosion
   c) Metamorphosis
   d) Organogenesis
688. The first postembryonic form of some insects (as the dragonfly) in which the larva is encased in a thin temporary membrane is called
   a) Nymph
   b) Pronymph
   c) Pharate pupa
   d) Subimgao

689. The developmental stage of a metamorphic insect prior to entering the pupal stage is

a) Prepupa b) Pronymph
c) Pharate pupa d) Subimgao

690. The uncovered pupae of butterflies are called

a) Cocoons b) Chrysalides/chrysalis
c) Puparium d) Nits

691. The hemimetabolous type of metamorphosis is also known as

a) Direct metamorphosis b) Incomplete metamorphosis
c) Direct metamorphosis d) Both a and b

692. The holometabolous type of metamorphosis is also known as

a) Indirect metamorphosis b) Complete metamorphosis
c) Direct metamorphosis d) Both a and b

693. In which of the insect order larvaes are aquatic and adults are terrestrial in habitat

a) Plecoptera b) Odonata
c) Dermaptera d) Both a and b

694. In which insect order the last immature stage possess functional wings

a) Diptera b) Dermaptera
c) Ephemeroptera d) Isoptera

695. The functional wings present in the last immature stage are termed

a) Imago b) Pronymph
c) Sub-imago d) Pharate

696. The type of larvae which emerge out in an early embryonic stage is

a) Protopod larva b) Oligopod larva
c) Polypod larva d) Apodous larva

697. In which type of larva segmentation, abdominal legs are present

a) Protopod larva b) Oligopod larva
c) Polypod larva d) Apodous larva

698. In which type of larva thoracic legs are less developed, abdominal legs are absent

a) Protopod larva b) Oligopod larva
c) Polypod larva d) Apodous larva

699. In which type of larva thoracic legs, abdominal legs are completely absent
   a) Protopod larva b) Oligopod larva
   c) Polypod larva d) Apodous larva
700. Protopod larvae are the characteristic features of
   a) Parasitic hymenoptera b) Diptera
   c) Thysanoptera d) Coleoptera
701. The examples of insect with characteristic Polypod larva are
   a) Lepidopteran larva b) Sawflies larva
   c) Scorpion flies larva d) All of the above
702. Campodeiform and scarabaeiform larva are the two types of
   a) Protopod larva b) Oligopod larva
   c) Polypod larva d) Apodous larva
703. Campodeiform type of larva is found in
   a) Neuroptera b) Strepsitera
   c) Trichoptera d) All of the above
704. C- shaped larvae also known as scarabaeiform type of larva are found in
   a) Scarabaeiodea b) Ptinidae
   c) Anobiidae d) All of the above
705. The type of Apodous larva with well sclerotized head capsule occurs in
   a) Eucephalous larva b) Hemicephalous larva
   c) Acephalous larva d) Both a and b
706. The type of Apodous larva with slight reduction in head capsule occurs in
   a) Eucephalous larva b) Hemicephalous larva
   c) Acephalous larva d) Both a and b
707. The type of Apodous larva with reduced head capsule occurs in
   a) Eucephalous larva b) Hemicephalous larva
   c) Acephalous larva d) Both a and b
708. The larvae of which insect family are commonly known as water tigers
   a) Gyrinidae b) Carabidae
   c) Hydrophilidae d) Dytiscidae
709. The phenomenon whereby a larval instar passes two distinct form during their life cycle is known as
   a) Metamorphosis b) Hypometamorphosis
   c) Hypermetamorphosis d) Pseudometamorphosis

710. The first instar larva in the beetle family Meloidae has three claws on each foot, and is therefore called a

a) Triungulin b) Inquiline

c) Wood worm d) Wireworm

711. The defensive secretion by the insect of Meloidae family is

a) Cantharidin b) Benzoquinone

c) Cardenolides d) Hydrogen poeroxide

712. Which of the following lepidopteran family, the phenomenon of Hypermetamorphosis is seen

a) Pyralidae b) Noctuidae

c) Gracilariidae d) Papilionidae

713. The larva of Elateridae are called

a) Red worms b) Meal worms

c) Wire worms d) Flat worms

714. The larva of Tenebrionidae are called

a) Red worms b) Meal worms

c) Wire worms d) Flat worms

715. The larva of chironomids are called

a) Red worms b) Meal worms

c) Blood worms d) Flat worms

716. Larva of Dobsonfly (Megaloptera) are called

a) Round worms b) Flat worms

c) Antlions d) Hellgrammites

717. According to Dyar's law the larval width of the head capsule in successive instar increases by a ratio of

a) 1.1 b) 1.2

c) 1.3 d) 1.4

718. The inactive stage in an insect life cycle is

a) Larva b) Pupa

c) Imago d) Sub-imago

719. The ability of an organism to survive in the dry state (Anhydrobiosis) is observed in larvae of

a) *Cluniomarinus* b) *Belgica antarctica*

c) *Polypedilum vanderplanki* d) *Musca domestica*

720. Jewel caterpillars belong to family

a) Zygaenidae b) Epipyropidae

c) Dalceridae d) Zygaenoidea

721. Which of the following family, the larva resembles birds dropping

a) Papilionidae b) Pieridae

c) Hespiridae d) Pterophoridae

722. The larvaes which are ectoparasites of mostly leafhopper and scale insects belongs to

a) Cyclotornidae b) Epipyropidae

c) Zygaenoidea d) Both a and b

723. In which of the following insect, the pupa are found active throughout the instar

a) Culicidae b) Chironomidae

c) Arctiidae d) Both a and b

724. Sclerotized and articulated mandibles which are used to escape from the cocoon are

a) Decticous pupa b) Adecticous pupa

c) Coarctate pupa d) Obtect pupa

725. Pupa with non-articulated mandibles occurs in

a) Adecticous pupa b) Exarate pupa

c) Obtect pupa d) All of the above

726. In which type of pupa the appendages lie freely without any secondary attachment

a) Decticous pupa b) Adecticous pupa

c) Exarate pupa d) Obtect pupa

727. In which type of pupa the appendages are pressed against its body and attached to substrate by the secretion of last larval moult

a) Decticous pupa b) Coarctate pupa

c) Exarate pupa d) Obtect pupa

728. The type of pupa enclosed in a puparium formed from last larval skin is

a) Decticous pupa b) Adecticous pupa

c) Coarctate pupa d) Obtect pupa

729. The sub order Terebrantia of Thysanoptera order have

a) Two pupal stages
b) Three pupal stages
c) Four pupal stages
d) Five pupal stages

730. The sub order Tubulifera of Thysanoptera order have

a) Two pupal stages
b) Three pupal stages
c) Four pupal stages
d) Five pupal stages

731. Larva of Chrysopidae are also known as

a) Antlion
b) Aphidlion
c) Mantidflies
d) Green lace wings

732. The first pupa like instar sometimes with external wing pads found in Thysanoptera and male Coccidae is known as

a) Prepupa
b) Propupa
c) Pharate pupa
d) All of the above

733. The pupa formed by the hardened last larval skin enclosing the pupa in some insects, especially higher Dipteran is known as

a) Prepupa
b) Propupa
c) Pharate pupa
d) Puparium

734. The formation of puparium is known as

a) Pupation
b) Perturbation
c) Pupating
d) Pupariation

735. The adult emergence from the pupa through an eversible cephalic sac called the Ptilinum which forces the opening of puparium is a characteristic feature of

a) Cyclorrhaphan
b) Isopteran
c) Lepidopteran
d) Tenthrinidae

736. In which of the following Hemipteran family, puparium is seen

a) Aleyrodidae
b) Psychidae
c) Fulgoridae
d) Membracidae

737. Pupal mating which is a form of reproductive strategy in which the adult male mates with a female pupa about to emerge, or with the newly moulted female is seen in

a) *Heliconius*
b) *Danaus*
c) *Biston*
d) *Helicoverpa*

738. In which of the Lepidopteran family, the pupa has multiple rows of spines near the anterior edge of the dorsum of abdominal segments

a) Noctuidae b) Saturinidae
c) Tortricidae d) Pyralidae

739. The waste product of pupal metabolism is known as

a) Osmeteria b) Meconium
c) Coremata d) Hairpencil

740. Pheromone signaling structures present in lepidopteran males are

a) Coremata b) Hairpencil
c) Osmeteria d) Both a and b

741. The developing of wings occur during

a) Pre embryonic growth b) Post embryonicgrowth
c) Embryonic growth d) None of the above

742. The mass of cells formed by holoblastic cleavage of egg is

a) Medulla b) Morula
c) Serosa d) Yolk

743. The nutritive or yolk of the egg in the cytoplasm is

a) Tonoplast b) Meroblast
c) Deutoplasm d) Serosa

744. The type of cleavage whereby the entire egg divides is

a) Holoblastic b) Mesoblastic
c) Meroblastic d) Hypoblastic

745. When the egg yolk attains a central position it is said to be

a) Centrolecithal b) Alecithal
c) Hololecithal d) Telolccithal

746. When the egg yolk lies at one end of the egg it is said to be

a) Centrolecithal b) Alecithal
c) Hololecithal d) Telolecithal

747. Insect eggs are

a) Centrolecithal b) Alecithal
c) Hololecithal d) Telolecithal

748. The parts of the cuticle discarded at moulting is

a) Frass b) Cuticula

c) Exuviae d) Tectocuticle

749. The skin irritation from contact with a vesicant toxin in the hemolymph in the genus *Paederus* of Coleopteran order is

a) Pederin b) Cantharidin

c) Quinone d) HCN

750. The agent which produces the skin irritation released by the Paederus insect is

a) *Fusarium* b) *Ralstonia*

c) *Phytophthora* d) *Pseudomonas*

751. The ability to produce natural antifreeze compounds is observed in

a) *Cucujus clavipes* b) *Stegobium paniceum*

c) *Lasioderma serricorne* d) *Lasioderma falli*

752. Which of the following insect is used as model organism for ethological and food safety

a) *Tenebrio molitor* b) *Tribolium confusum*

c) *Tribolium destructor* d) *Tribolium castaneum*

753. The phenomenon of Thanatosis is observed in

a) *Callosobruchus maculatus* b) *Callosobruchus chinensis*

c) *Callosobruchus phaseoli* d) All of the above

754. The phenomenon of Polyphenism is seen in

a) *Biston betularia* b) *Parnassiusapollo*

c) *Limenitis procris* d) *Eurema hecabe*

755. Batesian and Müllerian mimicry complexes are commonly found in

a) Coleoptera b) Diptera

c) Lepidoptera d) Hymenoptera

11

Registered Insecticides / Pesticides

Insecticides/Pesticides Registered under section 9(3) of the Insecticides Act, 1968 for use in the Country:
(As on 31/12/2018)

S.No.	Name of the Pesticide
1.	2,4-D Amine salt
2.	2,4-Dichlorophenoxy Acetic Acid
3.	Abamectin*
4.	Acephate
5.	Acetamiprid
6.	Afidopyropen
7.	Alachlor
8.	Allethrin
9.	Alphacypermethrin
10.	Alphanaphthyl Acetic Acid
11.	Aluminium Phosphide
12.	Ametroctradin
13.	Ametryn
14.	*Ampelomyces quisqualis*
15.	Anilophos
16.	Atrazine
17.	Aureofungin
18.	Azadirachtin (Neem Products)
19.	Azimsulfuron
20.	Azoxystrobin
21.	*Bacillus sphaericus*
22.	*Bacillus subtilis*
23.	*Bacillus thuringiensis* var. *galleriae*

24. *Bacillus thuringiensis* var. *israelensis*
25. *Bacillus thuringiensis* var. *kurstaki*
26. Barium Carbonate
27. *Beauveria bassiana*
28. Benalaxyl
29. Benalaxyl M
30. Bendiocarb
31. Benfuracarb
32. Bensulfuron Methyl
33. Bentazone
34. Beta Cyfluthrin
35. Bifenazate
36. Bifenthrin
37. Bispyribac Sodium
38. Bitertanol
39. Boscalid+ Pyraclostrobin
40. Bromadiolone
41. Buprofezin
42. Butachlor
43. Captan
44. Carbendazim
45. Carbofuran
46. Carbosulfan
47. Carboxin
48. Carfentrazone Ethyl
49. Carpropamid
50. Cartap Hydrochloride
51. Chlorantraniliprole
52. Chlorfenopyr
53. Chlorfluazuron
54. Chlorimuron ethyl
55. Chlormequat Chloride (CCC)
56. Chlorothalonil
57. Chlorpropham
58. Chlorpyrifes
59. Chlorpyrifes Methyl

60. Chromafenozide
61. Cinmethylene
62. Clodinafop-propargyl
63. Clodinafop-propargyl+Sodiumacifluorfen
64. Clomazone
65. Clothianidin
66. Copper Hydroxide
67. Copper Oxychloride
68. Copper Sulphate
69. Coumachlor
70. Coumatetralyl
71. Cuprous Oxide
72. Cyantraniliprole
73. Cyazofamid
74. Cyflumetofen
75. Cyfluthrin
76. Cyhalofop-butyl
77. Cymoxanil
78. Cypermethrin
79. Cyphenothrin
80. Dazomet
81. Deltamethrin (Decamethrin)
82. Diafenthiuron
83. Dichloro Diphenyl Trichloroethane (DDT)
84. Dichloropropene and Dichloropropane mixture (DD mixture)
85. Dichlorvos (DDVP)
86. Diclofop-Methyl
87. Diclosulam
88. Dicofol
89. Difenoconazole
90. Diflubenzuron
91. Dimethoate
92. Dimethomorph
93. Dinocap
94. Dinotefuron
95. Dithianon

96. Diuron
97. Dodine
98. D-trans Allethrin
99. Edifenphos
100. Emamectin Benzoate
101. Epoxyconazole
102. Ethephon
103. Ethion
104. Ethiprole
105. Ethofenprox (Etofenprox)
106. Ethoxysulfuron
107. Ethylene Dichloride and Carbon Tetrachloride mixture (EDCT Mixture 3:1)
108. Etoxazole(FI)
109. Famoxadone
110. Fenamidone
111. Fenazaquin
112. Fenitrothion
113. Fenobucarb (BPMC)
114. Fenoxaprop-p-ethyl
115. Fenpropathrin
116. Fenpyroximate
117. Fenvalerate
118. Fipronil
119. Flonicamid
120. Fluazifop-p-butyl
121. Flubendiamide
122. Flucetosulfuron
123. Fluchloralin
124. Fluensulfone 47% TC (MUP) (FI)
125. Flufenacet
126. Flufenoxuron
127. Flufenzine
128. Flumioxazin
129. Fluopicolide
130. Fluopyram and its metabolite

131. Flupyradifurone
132. Flusilazole
133. Fluvalinate
134. Fluxapyroxad
135. Fomesafen
136. Forchlorfenuron
137. Fosetyl-Al
138. Gibberellic Acid
139. Glufosinate Ammonium
140. Glyphosate
141. Haloxyfop-R-methyl 10.55%.EC(FI)
142. Helosulfuron methyl
143. Hexaconazole
144. Hexazinone
145. Hexythiazox
146. Hydrogen Cyanamide
147. Imazamox
148. Imazethapyr
149. Imidacloprid
150. Imiprothrin
151. Indoxacarb
152. Indaziflam+ Glyphosate ammonium (FI)
153. Iprobenfos (Kitazin)
154. Iprodione
155. Iprovalicarb
156. Isoprothiolane
157. Isoproturon
158. Kasugamycin
159. Kresoxim Mcthyl
160. Lambdacyhalothrin
161. Lime Sulphur
162. Lufenuron
163. Magnesium Phosphide Plates
164. Malathion
165. Mancozeb
166. Mandipropamid

167. Mepiquate Chloride
168. Meptyldiinocop
169. Mesosulfuron Methyl + Iodosulfuron Methyl Sodium
170. Metaflumizone
171. Metalaxyl
172. Metalaxyl-M
173. Metaldehyde
174. *Metarhizium anisopliae*
175. Methabenzthiazuron
176. Methomyl
177. Methyl Bromide
178. Methyl Chlorophenoxy Acetic Acid (MCPA)
179. Metiram
180. Metofluthrin
181. Metolachlor
182. Metrafenone
183. Metribuzin
184. Metsulfuron Methyl
185. Milbemectin
186. Monocrotophos
187. Myclobutanil
188. Novaluron
189. Nuclear polyhyderosis virus of *Helicoverpa armigera*
190. Nuclear polyhyderosis virus of *Spodoptera litura*
191. Orthosulfamuron
192. Oxadiargyl
193. Oxadiazon
194. Oxathiapiprolin
195. Oxycarboxin
196. Oxydemeton-Methyl
197. Oxyfluorfen
198. Paclobutrazol
199. Paraquat dichloride
200. Penoxasulam
201. Penconazole
202. Pencycuron

203. Pendimethalin
204. Penflufen
205. Penoxsulam
206. Permethrin
207. Phenthoate
208. Phorate
209. Phosalone
210. Phosphamidon
211. Picoxystrobin
212. Pinoxaden
213. Prallethrin
214. Pretilachlor
215. Primiphos-methyl
216. Profenophos
217. Prohexadione Calcium
218. Propamocarb hydrochloride
219. Propanil
220. Propaquizafop
221. Propergite
222. Propetamphos
223. Propiconazole
224. Propineb
225. Propoxur
226. *Pseudomonas fluorescens*
227. Pymetrozin (FI), TIM
228. Pyraclostrobin
229. Pyraclostrobin + Epoxiconazole
230. Pyrazosulfuron ethyl
231. Pyrethrin (pyrethrum)
232. Pyridalyl
233. Pyriproxyfen
234. Pyrithiobac sodium
235. Quinalphos
236. Quizalofop ethyl
237. Quizalofop-P-tefuryl
238. S-bioallethrin

239. Sodium paranitrophinolate
240. Spinetoram
241. Spirotetramat
242. Spinosad
243. Spiromesifen
244. Streptomycin + Tetracycline
245. Sulfentrazone
246. sulfosulfuron
247. Sulfoxaflor
248. sulphur
249. Tebuconazole
250. Tembotrione
251. Temephos
252. Tetraconazole (FI)
253. Thiacloprid
254. Thifluzamide
255. Thiobencarb (Benthiocarb)
256. Thiocyclam Hydrogen oxalate
257. Thiodicarb
258. Thiomethoxain
259. Thiophanate-Methyl
260. Thiram
261. Tolfenpyrad
262. Topramezone
263. Transfluthrin
264. Triacontanol
265. Triadimefon
266. Triafamone (Triafamone 20% w/w + Ethoxysulfuron 10% WG % w/w SC FI)
267. Triallate
268. Triasulfuron
269. Triazophos
270. Trichlorfon
271. *Trichoderma harzianum*
272. *Trichoderma viride*
273. Tricyclazole

274. Triflumezopyrim
275. Trifloxistrobin
276. Trifluralin
277. Validamycin
278. *Verticillium lecanii*
279. Zinc Phosphide
280. Zineb
281. Ziram

Source: DPPQS (Direcgtorcte of Plant Protection, Quarantine & Storage)

12

Model Question Paper (IARI PhD)

PART – I: (General Agriculture)

Multiple choice questions (No. 1 to 30)

1. Who is the present Chairman of Protection of Plant Varieties and Farmers' Right Authority (PPV&FRA)?

   a) Dr. R.R. Hanchinal　　b) Dr. P.L. Gautam
   c) Dr. S. Nagarajan　　d) Dr. Swapan K. Datta

2. Which among the following is another name for vitamin B_{12}?

   a) Niacin　　b) Pyridoxal phosphate
   c) Cobalamin　　d) Riboflavin

3. The largest share in India's farm export earnings in the year 2011-12 was from

   a) Basmati rice　　b) Non-basmati rice
   c) Sugar　　d) Guar gum

4. The National Bureau of Agriculturally Important Insects was established by ICAR in ________, was earlier known as _______.

   a) Bangalore; PDBC
   b) New Delhi; National Pusa Collection
   c) Ranchi; Indian Lac Research Institute
   d) New Delhi; NCIPM

5. The most important sucking pests of cotton and rice are respectively

   a) *Nilaparvata lugens* and *Aphis gossypii*
   b) *Aphis gossypii* and *Thrips oryzae*
   c) *Amrasca biguttula biguttula*and *Scirtothrips dorsalis*
   d) *Thrips gossypii* and *Orseolia oryzae*

6. Which of the following microorganism causes fatal poisoning in canned fruits and vegetables?

   a) *Aspergillus flavus*　　b) *Penicilliumdigitatum*
   c) *Clostridium botulinum*　　d) *Rhizoctonia solani*

7. The cause of the great Bengal Famine was
   a) Blast of rice b) Brown spot of rice
   c) Rust of wheat d) Karnal bunt of wheat
8. Actinomycetes belong to
   a) The fungi b) Eukaryote
   c) *Mycelia sterilia* d) None of the above
9. A virus-free clone from a virus infected plant can be obtained by
   a) Cotyledonary leaf culture b) Axenic culture
   c) Stem culture d) Meristem tip culture
10. Which of the following is not an objective of the National Food Security Mission?
   a) Sustainable increase in production of rice, wheat and pulses
   b) Restoring soil fertility and productivity at individual farm level
   c) Promoting use of bio-pesticides and organic fertilizers
   d) Creation of employment opportunities
11. Agmarknet, a portal for the dissemination of agricultural marketing information, is a joint endeavour of
   a) DMI and NIC
   b) DMI and Ministry of Agriculture
   c) NIC and Ministry of Agriculture
   d) DMI and Directorate of Economics and Statistics
12. The share of agriculture and allied activities in India's GDP at constant prices in 2011-12 was
   a) 14.1% b) 14.7%
   c) 15.6% d) 17.0%
13. The average size of land holding in India according to Agricultural Census 2005-06 is
   a) 0.38 ha b) 1.23 ha
   c) 1.49 ha d) 1.70 ha
14. 'Farmers First' concept was proposed by
   a) Paul Leagans b) Neils Rolling
   c) Robert Chamber d) Indira Gandhi

15. In the year 2012, GM crops were cultivated in an area of
    a) 150 million hectare in 18 countries
    b) 170 million hectare in 28 countries
    c) 200 million hectare in 18 countries
    d) 1.70 million hectare in 28 countries
16. The broad-spectrum systematic herbicide glyphosate kills the weeds by inhibiting the biosynthesis of
    a) Phenylalanine b) Alanine
    c) Glutamine d) Cysteine
17. At harvest, the above ground straw (leaf, sheath and stem) weight and grain weight of paddy crop are 5.5 and 4.5 tonnes per hectare, respectively. What is the harvest index of paddy?
    a) 45% b) 50%
    c) 55% d) 100%
18. Crossing over between non-sister chromatids of homologous chromosomes takes place during
    a) Leptotene b) Pachytene
    c) Diplotene d) Zygotene
19. The term 'Heterosis' was coined by
    a) G.H. Shull b) W. Bateson
    c) T.H. Morgan d) E.M. East
20. When a transgenic plant is crossed with a non-transgenic, what would be the zygosity status of the F_1 plant?
    a) Homozygous b) Heterozygous
    c) Hemizygous d) Nullizygous
21. The highest per capita consumption of flowers in the world is in
    a) The USA b) India
    c) Switzerland d) The Netherlands
22. Which of the following is a very rich source of betalain pigment?
    a) Radish b) Beet root
    c) Carrot d) Red cabbage
23. Dog ridge is
    a) Salt tolerant rootstocks of mango
    b) Salt tolerant rootstocks of guava
    c) Salt tolerant rootstocks of grape
    d) Salt tolerant rootstocks of citrus

24. Which of the following micronutrients are most widely deficient in Indian soils?
    - a) Zinc and boron
    - b) Zinc and iron
    - c) Zinc and manganese
    - d) Zinc and copper
25. Which of the following fertilizers is not produced in India?
    - a) DAP
    - b) Urea
    - c) Muriate of potash
    - d) TSP
26. What is the estimated extent of salt affected soils in India?
    - a) 5.42 mha
    - b) 7.42 mha
    - c) 11.42 mha
    - d) 17.42 mha
27. Which of the following is not a feature of watershed?
    - a) Hydrological unit
    - b) Biophysical unit
    - c) Socio-economic unit
    - d) Production unit
28. Correlation coefficient 'r' lies between
    - a) 0 and 1
    - b) -1 and 1
    - c) -1 and 0
    - d) 0 and -1
29. For the data 1, -2, 4, geometric mean is
    - a) 2
    - b) 4
    - c) -3
    - d) -2
30. The relationship between Arithmetic mean (A), Harmonic mean (H) and Geometric mean (G) is
    - a) $G^2=\underline{AH}$
    - b) G= /A+H
    - c) $H^2=GA$
    - d) $A^2=GH$

Multiple choice questions (No. 31 to 130).

Choose the correct answer (*a, b, c* or *d*) and enter your choice in the circle (by shading with a pencil) on the OMR - answer sheet as per the instructions given on the answer sheet.

31. Insect groups hitherto not known from India
    - a) Sialoidea
    - b) Zoraptera and Sialoidea
    - c) Grylloblattodea, Zoraptera,Mantophasmatodea
    - d) Raphidoidea and Embioptera

32. In India, the first Entomologist to be posted as "Entomologist to the Government of India" during the British Colonial regime is

a) Dr. D.G. Koenig
b) Lionel de Nice Velle
c) Maxwell Lefroy
d) Sir Ronald Ross

33. Avermectins are

a) Octopamine receptor agonists
b) Chloride channel activators
c) Nicotine acetylcholine receptor agonist
d) Sodium channel modulator

34. One of the serious endoparasite of mulberry silkworm is

a) Indian uzifly *Exorista sorbillans* of Tachinidae
b) *Exorista bombycis* of Tachinidae, Diptera
c) *Exorista* spp. of Eulophidae of Hymenoptera
d) *Exorista sorbillans* (Tachinidae: Diptera) onthe pupae

35. Mango stem borer complex belonging to the genus *Batocera* in India and adjacent S.E. Asian countries comprises of

a) Only *rufomaculata* DeGeer
b) *rufomaculata*, *rubus*, *royilei* and *numetor*
c) Only *rubus* (Linnaeus)
d) Only *rufomaculata* (DeGeer)

36. Inmulberry silkworm, the life cycle is completed in

a) Larvae with four moults and in 6 to 8 weeks
b) Larvae with six moults and in 6 to 8 weeks
c) Pupal period of 10-12 days and in 4 to 6 weeks
d) Larvae with eight moults and in 8 to 10 weeks

37. An insecticide of new chemical class of spirocyclic phenyl substituted tetronic acid and active against whiteflies and spider mites is

a) Imidacloprid
b) Spinosad
c) Spirodiclofen
d) Spiromesifen

38. Of the following, an insect demonstrated to show a hygrokinetic response with increasedactivity in moist air compared with dry air is

a) Wireworms
b) Locust
c) Hydrophilid beetle
d) Ephemeropteran nymph

39. A complex form of phototaxis occurring in a moving insect and receiving a constant visual stimulus is called as

a) Kinesis
b) Menotaxis
c) Photokinesis
d) Mechanophototaxis

40. Indian Tea Association started the entomological investigations in 1905 at

a) Tocklai (Assam)
b) UPASI, Valparai (Tamil Nadu)
c) Kannykorien (Assam)
d) Kannykorien (Tamil Nadu)

41. Androconia found in wings of certain insects are

a) Specialized scales
b) Bristles
c) Spots
d) Specialized glands

42. DDT was synthesized by

a) Zeidler, 1874
b) Paul Muller, 1939
c) Schrader, 1941
d) Aucante, 1954

43. In honey bee, brushes of stiff hairs are present in the inner surface of

a) Basitarsus of all the three castes and these are useful for cleaning of all the parts and materials
b) Foretarsus of all the three castes and these are useful for cleaning of all the parts and materials
c) Tibiae of all the legs in workers and these are useful in collecting pollen
d) Abdominal segments 4 to 7 for the purpose of wax diffusion

44. Caudal breathing tube is the characteristic feature of

a) Notonectidae
b) Nepidae
c) Corixidae
d) Naucoridae

45. Cells of haemolymph that take up foreign chemicals of high molecular weight

a) Nephrocytes
b) Oenocytes
c) Phagocytes
d) Nidi cells

46. Bald brood is a diseased condition in honey bees associated with infection of

a) Greater wax moth *Achroia grisella*
b) Lesser wax moth *Galleria mellonella*
c) Greater wax moth *Achroia innotata lankella*
d) Lesser wax moth *Achroia grisella*

47. The scientific name of the bee hunter wasp is
    a) *Palarus orientalis* b) *Vespa cincta*
    c) *Vespa tropica haematoides* d) *Vespa orientalis*
48. Apneumone is produced by
    a) Host plant b) Host insect
    c) Dead organic matter d) Parasitoid
49. Invaginations of integument in the head of an insect form
    a) Cibarium b) Tentorium
    c) Salivarium d) Atrium
50. Haluthonde is a disease caused by a
    a) Rickettsiae in *Apis cerana indica*
    b) Microsporidian in *Apis mellifera*
    c) Nuclear polyhedrosis virus, also called Grasserie
    d) Desonucleosis cytoplasmic polyhedrosis, also called as Grasserie
51. Johnston's organ is located in most of Pterygotes in
    a) Second antennal segment b) Legs
    c) Wings d) Third antennal segment
52. Government of India through its GEAC approved Bt cotton hybrids for commercialcultivation in India for the first time in
    a) March 2002, under MOEF
    b) March 2002, under Ministry of Agriculture
    c) March 2003, under Ministry of Agriculture
    d) March 2006, under MOEF
53. Kusmi and rangeeni cultures of lac insects are produced in the host plants
    a) *Shorea oleosa* and *Zizyphus mauritiana*,respectively
    b) *Zizyphus mauritiana* and *Shorea oleosa*,respectively
    c) *Shorea oleosa* and *Butea monosperma*,respectively
    d) *Shorea robusta* and *Zizyphus mauritiana*,respectively
54. In case of lac insects, the host species for summer sustainability are
    a) Kusum for kusmi summer crop and Palas for rangeeni summer crop
    b) Kusum for kusmi summer crop and *Ficus* spp. for rangeeni summer crop
    c) *Albizia lucida* for kusmi summer crop andPalas for rangeeni summer crop
    d) Palas for kusmi summer crop and Kusum for rangeeni summer crop

55. Of the following, the pest which feeds on the growing points of plants and thus retard thegrowth is

a) *Syllepta derogata* on cotton

b) *Maruca vitrata* on red gram

c) *Scelodonta strigicollis* on grapevine

d) *Helicoverpa armigera* on chickpea

56. Speciationwithout geographic isolation results in

a) Apomictic species b) Parapatric species

c) Allopatric species d) Palaeo species

57. In haemocoelic insemination, a special internal pouch for receiving sperms found infemale is known as

a) Mesospermalege b) Spermatophore

c) Spermatheca d) Bursa copulatrix

58. The fruit rust thrips infesting banana is

a) *Panchetothrips indicus*

b) *Helionothrips kadaliphilus*

c) *Thrips florum*

d) *Chaetanaphothrips signipennis*

59. Water dispersible granules differ from pelleted insecticides and are

a) having particle size in a range of 0.25 mm to 2.38 mm in dia (usually 250-1250 microns)

b) being impregnated or fused with toxicant and are flowables and without finely divided solids

c) being dry flowables and dry formulations of particle size 0.25 mm to 2.38 mm, with finely divided solids that combine with suspending and dispersing agents

d) being dry flowables and dry formulations, with finely divided solids that do not combine with suspending and dispersing agents

60. Lower and upper developmental thresholds for a species is respectively

a) The temperature below which development stops and the temperature at which the rate of growth or development begins to decrease

b) The temperature below which development stops and the temperature above which the rate of growth begins to decrease

c) The temperature above which growth stops and the temperature below which the rate of growth begins to increase

d) The temperature above which growth starts and the temperature above which the rate of growth begins to decrease

61. Niche of an organism refers to

a) Physiological position
b) Physical position
c) Functional position
d) Feeding position

62. *Sitophilus oryzae* lays eggs on

a) Grain
b) Frass
c) Gunny bags
d) Walls of container

63. In Embioptera, silk glands are situated on

a) Mid-tarsus
b) Hind-tarsus
c) Fore-tarsus
d) Abdomen

64. Grassy stunt virus disease of rice is transmitted by

a) Gall midge
b) Green leaf hopper
c) Brown plant hopper
d) White backed plant hopper

65. Indian Institute of Natural Resins and Gums was formerly known as, and belongs to

a) Resin Research Institute, CSIR
b) Indian Lac Research Institute, ICAR
c) Indian Gum Research Institute, ICAR
d) Gum & Resin Institute, CSIR

66. Japan has lifted in 2006, its two decade old ban on import of Indian mangoes on thecondition that the fruits are subjected to

a) Hot water treatment
b) Irradiation
c) Vapour heat treatment
d) Fumigation

67. Various parts of the body increase in linear dimensions by a ratio, which is constant for a species is known as

a) Pzibram law
b) Bioclimatic law
c) Dyar's law
d) Gaia hypothesis

68. Which part of an insect moult does not occur during apolysis?

a) Formation of new epicuticle
b) Resorption of old endocuticle
c) Formation of new quinone cross-linkages
d) Activation of moulting fluid

69. During the process of light production, luciferin is converted to
 a) Adenyl oxyluciferin b) Adenyl epoxyluciferin
 c) Adenyl hydroxyluciferin d) Adenyl carboxyluciferin
70. Corium is a part of
 a) Elytra b) Tegmen
 c) Hemilytra d) Forewing
71. Spiracles are absent in
 a) Cecidomyid larvae b) Mycetophilid larvae
 c) Chironomid larvae d) Psycodid larvae
72. Internally foregut is lined with
 a) Taenidia b) Peritrophic membrane
 c) Epithelial membrane d) Intima
73. Number of abdominal legs in the mustard saw fly larvae is
 a) 2 pairs b) 3 pairs
 c) 8 pairs d) 10 pairs
74. Of the following, one pest of apple is
 a) *Sylepta lunalis* b) *Cydia hemidoxa*
 c) *Nodostoma subcostatum* d) *Eriosoma lanigerum*
75. Nozzle in which spray fluid is fed into air stream within the nozzle and splitted into spray droplets by the velocity of air is called
 a) Disc nozzle b) Annular nozzle
 c) Shear nozzle d) Low volume nozzle
76. Granular or pelleted insecticides have generally a size range from
 a) 0.25 to 2.38 mm or usually 250-1250 microns of which a range of 100-300 microns are referred to as microgranules
 b) 1250-2250 microns of which the range >300 are referred to as granules
 c) 100-300 microns of which the range <300 are referred to as granules
 d) 1250-2250 microns and those with the range of 100-300 are referred to as microgranules
77. Hypostomal bridge is part of the
 a) Sutures of frons and it is present in Hymenoptera and Coleoptera
 b) Sclerites of head and it is found in Diptera
 c) Sclerites of protocephalon and is retained in the gnathocephalon of all insects
 d) Sclerites of head only, in its posterior aspect where it is called epistomal suture

78. In storage godowns, where curative methods are more important, the 'minifume' is

a) EDCT used at 24-32 kg/100 m^3

b) Ethylene dibromide developed as tablets and at a concentration of 56% and of 12 g

c) Ethylene dibromide developed as tablets by CFTRI, Mysore

d) Methyl bromide developed as tablets by CFTRI, Mysore at 2.5 to 3 kg/100 m^3

79. The formulation with WDL means and it is used when

a) Water diluted liquid, used when the active ingredient is insoluble in either water or organic solvents

b) Water dispersible liquid, used when the active ingredient is soluble in organic solvents and milled with a solid carrier

c) Water dispensable liquid, used when the active ingredient is insoluble in organic solvents and milled with a solid carrier

d) Water dispersible liquid, used when the active ingredient is insoluble in either water or organic solvents

80. Blister beetle and its product used in medicine are as follows

a) *Mylabris* spp. and Cantharidin, and it iscontained in the acrid oil excreted through the openings at the apex of legs

b) *Meloe* and allantoin, and it is contained in theacrid oil excreted through the openings at the apex of legs

c) *Mylabris* spp. and allantoin, which is usedexternally as a vesiccant and counter irritant

d) *Meloe* and *Mylabris* spp., and it is containedin the acrid oil excreted through the openings at the apex of legs

81. Froghoppers belong to the family

a) Cercopsidae b) Amphibiidae

c) Cicadellidae d) Cercopidae

82. Closely packed layer formed by tracheae at the back of the eye in many Lepidoptera is called

a) Semper's cells b) Rhabdom

c) Retinula d) Tapetum

83. Successful biological control of cassava mealy bug in Africa was contribution by

a) Ray F. Smith b) Hans Herren

c) C. Bushland d) Sweetman

84. The layer of egg that is formed when the egg is in the ovary itself

a) Serosa
b) Cement layer
c) Chorion
d) Yolk

85. Uncoupling of oxidative phosphorylation is the principal mode of action in the recently developed

a) Pyrrole insecticides
b) Organophosphates
c) Synthetic pyrethroids
d) Azadirachtin based insecticides

86. Spinosad as an insecticide is

a) Derived from soil microorganisms and it is an acetylcholine receptor agonist similar to neonicotinoids
b) An acetylcholine receptor agonist similar to neonicotinoids in the end result, but somewhat different from them
c) A chloride channel inhibiting agent, affecting the GABA receptor
d) A GABA gated chloride channel agonist and nerve impulses are then unable to travel down the chloride channel

87. Adikokinetic hormone is

a) Peptide hormone
b) Steroid activated by a terpenoid
c) Sesquiterpenoid activated by a peptide
d) Peptide hormone induced by amino peptidase

88. Hexythiazox is a

a) PTTH synthesis interfering non-specific growth regulator and kills the eggs before hatching
b) Non-specific growth regulator affecting the eggs before hatching and killing the adults too
c) Non-specific growth regulator, kills the eggs before mites hatch, also immature and adult mites are not killed
d) Interferes with PTTH synthesis and production of ecdysone and thus a non-specific growth regulator

89. Of the following, example of quinoxalineorganothiophosphate is

a) Pirimiphos methyl
b) Diazinon
c) Fenitrothion
d) Quinalphos

90. Nontoxic to bees and decompose in water to yield fumigant vapour is characteristic of

a) Propoxur
b) All dithiocarbamates
c) Dazomet
d) Dazomet (propoxur RTU)

91. Which of the following is correct?
   a) 1914 - Melander - San Jose Scale - lime sulphur
   b) 1914 - Melander - San Jose Scale - hydrogen cyanide
   c) 1914 - Melander - California Red Scale on Citrus - lime sulphur
   d) 1916 - Quayle – California Red Scale - lime sulphur

92. The following are all examples of insecticides banned in India
   a) Aldrin, aldicarb, endrin, propoxur
   b) Aldrin, benzene hexachloride, dieldrin, ethylene dibromide
   c) Aldrin, dieldrin, endrin, propoxur
   d) Benzene hexachloride, endrin, dieldrin

93. Some of the insecticides of restricted use in India are
   a) Monocrotophos (banned for agriculture), endosulfan (banned in Kerala state)
   b) Fenthion (banned for vegetables), endosulfan (banned in Kerala state)
   c) Monocrotophos (banned for vegetables), endosulfan (banned in Kerala state)
   d) Fenthion and diazinon (banned for vegetables)

94. *Acerophaguspapayae*is a parasite deployedfor the biological control of
   a) lac insect parasites as hyperparasite
   b) Papaya mealy bug *Paracoccus marginatus*
   c) Papaya mealy bug *Phenacoccus papayae*
   d) Apple wooly aphid *Icerya purchasi*

95. Examples of internal feeder + external feeder, respectively, occurring on stored cereals
   a) *Sitophilus oryzae* + *Sitotroga cerealella*
   b) *Sitophilus oryzae* + *Lasioderma serricorne*
   c) *Callosobruchus oryzae* + *Callosobruchus chinensis*
   d) *Rhyzoperthadominica*+ *Phthorimaea opercullella*

96. *Glyphodespulveruntalis*is a pest of
   a) Cucurbits, bores stems and is Crambidae: Lepidoptera
   b) Mulberry, bores stems and is Crambidae: Lepidoptera
   c) Cucurbits, folds leaves and buds and is a Glyphidae: Lepidoptera
   d) Mulberry, folds leaves and buds, and is a Crambidae: Lepidoptera

97. The banana rhizome weevil is

a) *Odoiporuslongicollis*, it oviposits in rootstock or leaf sheath, feeds on rhizomes

b) *Odoiporuslongicollis*, it ovipositsinpseudostem, feeds on pseudostem

c) *Cosmopolites sordidus*, it oviposits on leafsheath and feeds on pseudostem

d) *Cosmopolites sordidus*, it oviposits in rootstock/leaf sheath just above ground level and feeds on rhizomes

98. TheNational Institute of Communicable Diseases is located at

a) Delhi and its old name is Malaria Institute of India

b) Kasauli and its old name is Malaria Survey of India

c) Tejpur and its old name is King Institute of Preventive Medicine

d) Kolkata and its old name is Institute of Nuclear Medicine and Allied Sciences

99. Of the following, who were both forest entomologists in India prior to independence?

a) M.L. Roonwal and E.P. Stebbing

b) A.D. Imms and E.P. Stebbing

c) R.N. Mathur and E.P. Stebbing

d) C.F.C. Beeson and R.N. Mathur

100. Of the following, who is associated with Arcana Entomologica?

a) Dr. J.G. Koenig b) Fabricius

c) Donovan d) Westwood

101. Of the following, with C=cost of control; P=marker value of crop; D_1 and D_2=damage functions, the EIL is given by the two equations, namely

a) EIL = C/PD_1; EIL = C/PD_2 b) EIL = C PD_1; EIL = C/PD_1

c) EIL = C PD_2; EIL = C/PD_2 d) EIL = C-PD_2; EIL = C/PD_1

102. The first BPH resistant variety of rice is

a) Mudgo and it is resistant to all populations of BPH

b) IR26 and it is resistant to all populations of BPH

c) Mudgo and it was not resistant to populations from India

d) Mudgo and it was resistant to populations from India

103. Grapevine stem borer and stem girdler are respectively

a) *Stheniasgrisator*(Cerambycidae), *Scelodonta strigicollis* (Chrysomelidae)

b) *Stheniasgrisator*and*Coelosternascabrator*(both Cerambycidae)

c) *Coelosternascabrator*and*Stheniasgrisator*(both Cerambycidae)

d) *Coelosternascabrator* (Chrysomelidae), *Stheniasgrisator* (Cerambycidae)

104. Of the following, the correct combination is

a) Potato - *Cylasformicarius* - Apionidae; Sweet potato - *Phthorimaea opercullella* - Gelechiidae

b) Sweet potato - *Omphisaanastomosalis* - Pyralidae; potato - *Phthorimaea opercullella* - Gelechiidae

c) Potato - *Epilachnavigintioctopunctata* - Coccinellidae; brinjal - *Epilachnavigintioctopunctata*- Coccinellidae

d) Sweet potato - *Epilachnavigintioctopunctata*- Coccinellidae; potato -*Aspidomorphamiliaris*- Chrysomelidae

105. Of the following, the singing insect is

a) *Cybisterjaponicus*

b) *Gryllotalpagryllotalpa*

c) Katydid *Gampsocleisgratiosa* Tang

d) Katydid *Gampsocleisgratiosa* Brunner

106. Of the following, example of a persistent pest is

a) Fruit borers on brinjal and okra

b) Aphids on mustard

c) Grasshoppers on rice

d) Scales and mealy bugs on horticultural crops

107. Systemsanalysis and modeling for prediction of damage and yield loss depends on the following

a) Qualitative techniques, understanding interactions, in components of cropping system, relationship with management practice

b) Qualitative and quantitative techniques, understanding interactions, in components of crop-pest system, relationship with management practice

c) Qualitative and quantitative techniques, understanding interactions, in components of crop-pest system, and relationship with management practice is not involved

d) Qualitative and quantitative techniques, understanding interactions, relationship with management practice but components of crop-pest system is not involved

108. FIK and RTU are respectively used with regard to

a) Flying insect killer sprays and ready to use aerosols

b) Fly instant killer aerosols and ready to use sprays

c) Flying insect killer aerosols and ready to use sprays

d) Flying insect killer aerosols and ready total use sprays

109. Removing of excess water from the extra cellular compartment in insect is known as

a) Diuresis b) Osmotic homeostasis

c) Osmoregulation d) Antidiuresis

110. Apple woolly aphis causes damage to

a) Root b) Stem

c) Root and stem d) Inflorescence

111. By addition of two toxic compounds, when combined toxicity is increased more thanthe combined effect, the effect is termed as

a) Synergism b) Antagonesim

c) Additive effect d) Potentiation

112. Insect order with panarpoid origin

a) Siphonaptera b) Odonata

c) Strepsiptera d) Siphunculata

113. When only last pair of abdominal spiracles are open, it is termed as

a) Peripneustic b) Propneustic

c) Amphipneustic d) Metapneustic

114. One Galleria unit represents the activity of JH which results in

a) Half of the (50%) treated insects retain pupal characters at the sight of treatment

b) Half of the (50%) treated insects retain larval characters at the sight of treatment

c) Complete (100%) of the treated insects retain pupal characters at the sight of treatment

d) Complete (100%) of the treated insects retain larval characters at the sight of treatment

115. Apterygotes that do not have entognathus mouth parts

a) Diplura b) Protura

c) Thysanura d) Collembola

116. Blood of insects differs from human blood due to presence of high amount of

a) Haemoglobin b) Pigments

c) Free amino acids d) Globulin proteins

117. Among the following, chitin synthesis inhibitor of plant origin is

a) Muscarone b) Plumbagin

c) Nuvaluron d) Canvanine

118. Ahormone mainly responsible for melanization of cuticle in Diptera is

a) Juvenile hormone b) Ecdysone

c) Proctolin d) Bursicon

119. The pure culture is called

a) Axenic b) Xenic

c) Synxenic d) Gnotobiotic

120. Prothoracic glands persist even in adults of

a) Cockroaches b) Beetles

c) Moths and butterflies d) Silverfish

121. "Probits" used in bioassay were first termed so by

a) Bliss b) Panse & Sukhatme

c) Finney d) Fisher

122. Rotenone interferes with electron transport between

a) Reduced cyt. a and cyt. a_3 b) Cyt. a and reduced NAD

c) Cyt. c and reduced FAD d) Cyt. a_3 and oxygen

123. Which of the following insects is a root borer of sugarcane?

a) *Chilo tumidicostalis* b) *Chilo infuscatellus*

c) *Raphimetopus ablutellus* d) *Emmalocera depressella*

124. Commercially used non-ester type JH mimic

a) Methoprene b) Hydroprene

c) Pyriproxyfen d) Flufenxuron

125. Regeneration of broken legs in nymphal stage is a characteristic feature of

a) Isoptera b) Orthoptera

c) Phasmida d) Lepidoptera

126. Insect classification presently followed is mostly based on the classification given by

a) Comstock and Comstock b) Handlirsch modified by Imms

c) Linnaeus d) Tillyard

127. 'Propodeum' in Hymenoptera is formed by the fusion of

a) Pro thorax and meta thorax b) Meta thorax and abdomen

c) Meso thorax and meta thorax d) Thorax and abdomen

128. Maximum Residual Limits of pesticides are fixed by the

a) Food and Drug Administration Authority

b) Codex Alimentarius Commission

c) Environmental Protection Agency

d) World Health Organization

129. In which one of the larvae of endopterygotes, compound eyes are noticed?

a) Lepidoptera b) Mecoptera

c) Hymenoptera d) Diptera

130. Snowdrop lectin, which is effective against aphids is obtained from

a) *Phaseolus vulgaris* b) *Galanthus nivalis*

c) *Lycopersicon peruvianum* d) *Solanum berthaltii*

13

Match the Appropriate Options

1. Match the sclerites of the head with their location

1. Vertex	i) Facial area below the vertex and above clypeus
2. Frons	ii) Top of the head between compound eyes
3. Clypeus	iii) Between occipital and post occipital suture
4. Gena	iv) Lateral cranial area behind the compound eyes
5. Occiput	v) Below the frons to which labrum is attached

2. Match the following antennae found in the insect

1. Setaceous (bristle like)	i) Odonata, Leafhopper, cockroach
2. Filliform (thread like)	ii) Grasshopper
3. Moniliform (beaded)	iii) Termites
4. Serrate (saw like)	iv) Buprestid Beetle
5. Unipectinate (comb like)	v) Sawfly

3. Match the following

1. Arachnida	i) First pair of claw as poison claw
2. Chilopoda	ii) Respiration by Booklungs
3. Diplopoda	iii) Cutaneous respiration
4. Onychophora	iv) Rolling as defense mechanism
5. Crustacean	v) Body region divided into two (Cephalothorax and abdomen)

4. Match the following antennae found in the insect

1. Bipectinate (double comb like)	i) Silkworm moth
2. Clavate (clubbed)	ii) Blister beetle
3. Capitate (knobbed)	iii) Butterfly
4. Lamellate (plate like)	iv) Lamellicorn beetle
5. Aristate	v) Housefly

5. Match the following antennae found in the insect

1. Stylate	i) Robberfly, Horse flies
2. Plumose (feathery)	ii) Male mosquito
3. Pilose (hairy)	iii) Female mosquito
4. Geniculate (elbowed)	iv) Ant, weevil and Honey bee
5. Flabellate (fan like)	v) Strepsiterans, Cedar beetles

6. Match the following mouthparts of an insect

1. Piercing and sucking	i) Honey bee
2. Chewing and lapping	ii) Bugs
3. Rasping and sucking	iii) Housefly
4. Sponging type	iv) Butterflies
5. Siphoning type	v) Thrips

7. Match the following types of insect legs

1. Ambulatorial	i) Forelegs of mole cricket
2. Cursorial	ii) Head louse (3 pairs of legs)
3. Saltatorial	iii) Hind legs of grasshopper
4. Scansorial	iv) Cockroach
5. Fossorial	v) Foreleg and middle leg of grasshopper

8. Match the following types of insect legs

1. Raptorial	i) Hind legs of bee
2. Natatorial	ii) Legs of Odonata
3. Sticking leg	iii) Housefly
4. Basket like leg	iv) Hind legs of water bug and water beetle
5. Foragial leg	v) Forelegs of Preying Mantids

9. Match the following legs of foragial type with their function

1. Eye brush (fore leg)	i) Corbicula (load pollen and Propolis)
2. Strigillis (fore leg)	ii) Load pollen in Corbicula
3. Pollen brush (fore leg)	iii) Collect pollen from the head and mouthparts
4. Pollen basket (hind leg)	iv) Antennae cleaner
5. Pollen packer (hind leg)	v) Clean the compound eyes

10. Match the following types of insect wing

1.	Tegmina	i)	Forewings of Hemipterans
2.	Elytra	ii)	Hind wings of Diptera
3.	Hemelytra	iii)	Thrips
4.	Halteres	iv)	Forewings of Coleopterans
5.	Fringed wings	v)	Forewings of Orthoptera

11. Match the following types of wing coupling found in insects

1.	Primitive type	i)	Butterflies
2.	Jugate wing coupling	ii)	Hymenoptera
3.	Frenate wing coupling	iii)	Micropterigidae
4.	JugoFrenate coupling	iv)	Fruit sucking moths and hawk moths
5.	Hamuli	v)	Hepialidae (Lepidoptera), Strepsiterans
6.	Amplexiform	vi)	Mecoptera, Trichoptera

12. Match the following abdominal appendages found in immature insects

1.	Filamentous gills (lateral gills)	i)	Naiads of Dragonfly
2.	Caudal gills (leaf gills)	ii)	Mosquito larva
3.	Rectal gills	iii)	Lepidopteran larva
4.	Anal papillae	iv)	Naiads of Damselfly
5.	Prolegs	v)	Naiads of Mayfly

13. Match the following diversity and form of cerci found in insect

1.	Long and many segmented	i)	Earwigs
2.	Long and unsegmented	ii)	Grasshopper
3.	Short and many segmented	iii)	Cockroach
4.	Short and unsegmented	iv)	Cricket
5.	Forceps like	v)	Mayfly

14. Match the different types of ovipositors found in insects

1.	Short and horny	i)	Short horned grasshopper
2.	Long and sword like	ii)	Katydids, long horned grasshopper
3.	Needle like	iii)	Cricket
4.	Ovipositor modified into sting	iv)	Worker of honey bee
5.	Pseudo ovipositor	v)	Fruitflies and Houseflies

15. Match the following sensory hairs present in an insect with their shape

1. Sensilla trichodea	i) Setae like
2. Sensilla chaetica	ii) Spine like
3. Sensilla basiconia	iii) Pegs or cone like
4. Sensilla placodea	iv) Plate like
5. Campaniform sensilla	v) Dome shaped

16. Match the following compound eyes with their examples

1. Eucone eyes	i) Thysanoptera, Orthoptera etc
2. Pseudocone eyes	ii) Nematoceran, Dermaptera,
3. Acone eyes	iii) Brachyceran and Cyclorhaphan
4. Exocone eyes	iv) Dermistidae, Elateridae, Cantharidae

17. Match the different respiratory system with their functioning spiracles found in an insect

1. Peripneustic	a) None of the spiracles are functioning
2. Amphipneustic	b) 9 spiracles open (Metathorax closed)
3. Propneustic	c) Prothoracic and Posterior abdominal spiracles open
4. Metapneustic	d) Last pair of abdominal spiracle open
5. Apneustic	e) Only prothoracic spiracles are open

18. Match the following respiratory gills along with the examples

1. Abdominal gills	a) Zygoptera nymphs
2. Thoracic gills	b) Caddis fly larva
3. Blood gills	c) Ephemeroptera
4. Lamellate and filamentous	d) Stonefly nymph
5. Caudal gills	e) *Chironomus*

19. Match the following abdominal legs with their example

1. Prolegs on 6 and 10th abdominal segment	a) Mustard Sawfly
2. Prolegs on 5, 6 and 10th abdominal segment	b) Dragonflies
3. Prolegs on 3, 4, 5, 6 and 10 abdominal segment	c) Cabbage looper
4. 8 pairs of Prolegs	d) Semiloopers
5. Prehensile legs	e) Tobacco caterpillar

20. Match the different types of reproduction with their examples

1. Ovoviviparity	a) Lepidoptera
2. Oviparity	b) *Glossina*
3. Adenotrophic viviparity	c) *Cimex*
4. Haemocoelous viviparity	d) Aphids
5. Pseudoplacental viviparity	e) *Miaster*

21. Match the different types of Parthenogenis with their example

1. Obligate	a) *Bombyx mori*
2. Apomictic (meiotic)	b) Aphids
3. Cyclic	c) Phasmids
4. Sporadic	d) Aphids
5. Apomictic (ameiotic)	e) Honey bee

22. Match the different parts of head with its description

1. Epistomal suture	a) transverse groove located behind the head
2. Hypostoma	b) depressions in the cranium wall
3. Hypostomal suture	c) sutures uniting the anterior ends of subgenae
4. Occipital suture	d) suture of subgenae lying posterior to mandible
5. Tentorial pits	e) part of cranium lying posterior to mandible

23. Match the different parts of head with its description

1. Epimeron	a) pleuron posterior to pleural suture
2. Notum	b) Tergum of thorasic segment
3. Alinotum	c) dorsal thorasic plate to which wings are attached
4. Phragmata	d) plate like apodemes
5. Scutellum	e) posterior portion of mesonotum/ metanotum of thorax

24. Match the different parts of wings with its description

1. Axillary plate	a) posterior sclerite of Odonata wing base
2. Costa	b) first vein of the wing
3. Cross vein	c) links the longitudinal veins
4. Frenulum	d) bristles in hind wings in contact with forewings
5. Sub costa	e) second vein of the wing

25. Match the terminology with their meaning

1. Dilator muscles	a) group of retinal cells
2. Arborization	b) one unit of a compound eye
3. Recurrent nerve	c) nerve extending posteriorly from frontal ganglion
4. Ommatidium	d) fine branches of axons or collaterals
5. Retinula	e) the muscles from body wall to alimentary canal

26. Match the following sensilla with its description

1. Sensilla opticum	a) photoreceptor of compound eye
2. Sensilla scolopophorym	b) sense organ with sense rod of scolops type
3. Sensilla squamiformium	c) scale like sensilla
4. Sensilla trichodeum	d) seta like sensilla
5. Sensilla placodeum	e) plate like sensilla

Appendices

Appendix I

Insecticides approved to control household pests in houses under the Insecticides Act, 1968

S.No.	Name of the Pesticide	Formulation
1.	Allethrin	0.2% Coil, 0.5% Coil, 4% Mat, 3.6% Liquid Vaporizer, 0.5% Aerosol
2.	Alphacypermethrin	0.5% chalk, 0.1% RTU, 5% WP
3.	Beta Cyfluthrin	2.45% SC
4.	Bifenthrin	0.05% Mosquito Coil
5.	Cyfluthrin	5% EW, 10% WP
6.	Cyfluthrin 0.025% + Transfluthrin 0.04% water based Aerosol	Aerosol
7.	Cypermethrin	3% Smoke generator, 1% chalk, 0.1% Aqueous
8.	Cypermethrin 0.11% + Pyrethrin 0.2% Aerosol	Aerosol
9.	Cyphenothrin	5% EC
10.	Cyphenothrin 0.3% + d-allethrin 0.2% Aerosol	Aerosol
11.	Deltamethrin	0.5% chalk, 1.25% ULV, 2.5% Flow, 0.5% tablet bait, 1% RTU
12.	Deltamethrin 2.5% + Allethrin 2.0% Liquid Concentrate	(to be used only by pest control operators, Government establishments, Government agencies for large scale disinfections)
13.	Diazinon	25% Micro Encapsulation
14.	Diflubenzuron	2% GR, 2% Tablet, 25% WP
15.	d-trans allethrin	2% Mat, 0.1% Coil
16.	Fenitrothion	2% Spray, 20% OL
17.	Fenthion	2% Spray
18.	Fipronil	0.05% Gel
19.	Forchlorfenuzon (CPPU)	0.1% Liquid
20.	Imidacloprid	2.15% Gel

21.	Lambda cyhalothrin	10% WP, 0.5% Chalk, 2.43% CS (Impregnated Bed Nets for vector control)
22.	Malathion	2% Spray, 5% Spray
23.	Metofluthrin	0.005% M.coil (6 Hrs.)
24.	Permethrin	5% SG
25.	Pirimiphosmethyl	1% Spray
26.	Prallethrin	0.05% Coil, 0.04% Coil, 0.5% Mat, 0.8% Mat, 1% red Mat, 1.2% Mat, 1.6% liquid Vaporizer, 0.8% Liquid Vaporizer, 19% VP, 2.4% LV
27.	Propetamphos	1% Spray
28.	Propoxur	20% EC, 1% Spray, 2% Aerosol, 2% Bait
29.	Pyrethrum	0.2% Aerosol
30.	S-Bioallethrin	2.4% Mat
31.	Transfluthrin	0.88% Liquid, 20% MV Gel(30 days mat tray), 1.6% LV, 1.2 % LV

As on 20th October 2015

Appendix II

Insecticides approved by the registration committee for protecting buildings from Termites

S.No.	Insecticide	Dosage
1.	Bifenthrin 2.5% EC	20 ml in 1 liter water
2.	Chlorpyriphos 50% EC , 20% EC	50 ml in 5 liter water
3.	Ethion 50% EC	50 ml in 5 liter water
4.	Imidacloprid 30.50% SC	10.5 ml in 5 liter water

As on 20th October 2015

Appendix III

Insecticides approved by the registrion committee for the Controls of Stored Grain Pests under the Insecticides Act, 1968

S.No.	Insecticide	Formulation
1.	*Aluminium Phosphide	56% m/m
2.	*Aluminium Phosphide	15% Tablet, 6% Tablet
3.	Aluminium Phosphide	77.5% GR.
4.	Deltamethrin	2.5% WP
5.	*Methyl Bromide Technical	Technical
6.	Methyl Bromide 98% + Chloropicrin 2% w/w	Fumigant

*To be used by Govt. approved agencies under expert supervision only

As on 20th October 2015

Appendix IV

Biocontrol laboratories in India

Type of Lab	Number of Labs
CIPMCs	35
SBCL (Grant in aid)	38
ICAR	49
SBCL	98
Private	141
Total	361

Appendix V

Pesticide testing Laboratories in State/UTs

S.No.	State/UT	No. of laboratories	Place
1.	Andhra Pradesh	7	Rajendra Nagar, Guntur, Anantapur, Tadepalligudem & Warangal, Vishakhapatnam and Kurnool
2.	Arunachal Pradesh	1	Naharlagun
3.	Assam	1	Guwahati
4.	Bihar	1	Patna
5.	Chhattisgarh	1	Raipur
6.	Gujarat	2	Junagarh&Gandhinagar
7.	Haryana	4	Karnal, Sirsa, Rohtak&Panchkula*
8.	Himachal Pradesh	1	Shimla
9.	Jammu & Kashmir	2	Srinagar & Jammu
10.	Karnataka	6	Bangalore, Bellary, Dharwad, Shimoga, Kotnoor & Mandya
11.	Kerala	1	Trivendrum
12.	Madhya Pradesh	1	Jabalpur
13.	Maharashtra	4	Pune, Amaravathi, Thane & Aurangabad
14.	Manipur	1	Mantripukhri
15.	Mizoram	1	Neihbawih
16.	Odisha	1	Bhubaneshwar
17.	Pondicherry	1	Pondicherry
18.	Punjab	3	Amritsar, Ludhiana &Bhatinda
19.	Rajasthan	6	Jaipur, Bikaner, Udaipur, Kota, Jodhpur & Sriganganagar
20.	Tamil Nadu	15	Coimbatore, Kovilpatti, Erode, Madurai, Trichy, Aduthrai, Salem, Cuddalore & Kanchipuram, Theni, Nagapattinam, Dharmpuri, Vellore, Sivaganga, Tirunelveli
21.	Tripura	1	Agartala

22.	Uttar Pradesh	4	Meerut, Lucknow (2 SPTLs) & Varanasi
23.	Uttarakhand	2	Rudrapur, Srinagar (Pauri Garhwal)
24.	West Bengal	1	Midnapore
Regional Pesticide Testing Laboratories			
1.	Kanpur	1	
2.	Chandigarh	1	
Central Insecticides Laboratory			
1.	Faridabad	1	

Appendix i-v *Source*: DPPQS

Answer Key

1.	A	37.	A	73.	A	109.	C	145.	B
2.	A	38.	A	74.	D	110.	A	146.	C
3.	D	39.	B	75.	A	111.	B	147.	A
4.	C	40.	D	76.	C	112.	A	148.	D
5.	B	41.	B	77.	A	113.	C	149.	D
6.	B	42.	A	78.	B	114.	D	150.	C
7.	C	43.	B	79.	A	115.	A	151.	A
8.	D	44.	C	80.	B	116.	A	152.	A
9.	B	45.	A	81.	A	117.	A	153.	A
10.	D	46.	B	82.	B	118.	A	154.	D
11.	C	47.	D	83.	C	119.	A	155.	C
12.	C	48.	A	84.	B	120.	D	156.	A
13.	C	49.	B	85.	B	121.	D	157.	D
14.	C	50.	A	86.	B	122.	C	158.	A
15.	A	51.	B	87.	A	123.	B	159.	A
16.	B	52.	C	88.	C	124.	A	160.	D
17.	D	53.	C	89.	B	125.	A	161.	A
18.	C	54.	B	90.	A	126.	A	162.	D
19.	A	55.	C	91.	D	127.	B	163.	C
20.	A	56.	C	92.	C	128.	C	164.	D
21.	D	57.	A	93.	C	129.	A	165.	A
22.	C	58.	B	94.	B	130.	A	166.	A
23.	A	59.	C	95.	A	131.	D	167.	B
24.	C	60.	B	96.	C	132.	A	168.	A
25.	A	61.	B	97.	C	133.	A	169.	A
26.	C	62.	A	98.	B	134.	C	170.	D
27.	B	63.	C	99.	C	135.	D	171.	A
28.	B	64.	B	100.	A	136.	D	172.	B
29.	D	65.	C	101.	A	137.	A	173.	C
30.	B	66.	C	102.	D	138.	D	174.	B
31.	A	67.	B	103.	B	139.	C	175.	D
32.	A	68.	B	104.	C	140.	D	176.	B
33.	B	69.	B	105.	C	141.	A	177.	B
34.	A	70.	A	106.	D	142.	A	178.	C
35.	B	71.	C	107.	B	143.	C	179.	C
36.	D	72.	D	108.	B	144.	D	180.	C

181. C
182. C
183. C
184. B
185. A
186. D
187. B
188. A
189. B
190. A
191. B
192. A
193. B
194. D
195. A
196. D
197. A
198. B
199. D
200. C
201. C
202. A
203. C
204. A
205. A
206. A
207. D
208. B
209. D
210. D
211. C
212. B
213. B
214. C
215. A
216. D
217. D
218. B
219. D
220. A
221. D
222. A
223. A
224. D
225. C
226. A
227. A
228. D
229. D
230. A
231. B
232. A
233. B
234. A
235. D
236. B
237. B
238. C
239. B
240. A
241. D
242. C
243. C
244. D
245. B
246. C
247. B
248. D
249. A
250. C
251. B
252. D
253. B
254. D
255. C
256. B
257. B
258. A
259. B
260. C
261. A
262. B
263. C
264. C
265. A
266. B
267. A
268. B
269. B
270. A
271. B
272. A
273. B
274. C
275. A
276. A
277. C
278. D
279. D
280. D
281. D
282. A
283. B
284. B
285. D
286. A
287. A
288. C
289. B
290. B
291. A
292. C
293. D
294. B
295. B
296. A
297. A
298. D
299. B
300. C
301. B
302. C
303. B
304. A
305. B
306. A
307. B
308. C
309. D
310. A
311. A
312. D
313. B
314. A
315. B
316. B
317. A
318. A
319. D
320. D
321. A
322. A
323. A
324. B
325. C
326. D
327. D
328. D
329. A
330. B
331. A
332. B
333. A
334. A
335. A
336. B
337. A
338. B
339. C
340. C
341. A
342. A
343. D
344. C
345. D
346. D
347. C
348. C
349. A
350. D
351. A
352. D
353. C
354. A
355. C
356. D
357. B
358. B
359. B
360. D
361. D
362. A
363. D
364. A
365. D
366. A
367. C
368. C
369. D
370. B
371. C
372. B
373. A
374. A
375. C
376. B
377. A
378. A
379. B
380. D
381. D
382. D
383. A
384. C
385. C
386. B
387. A
388. A
389. D
390. A
391. A
392. C
393. D
394. B
395. D
396. C
397. A
398. D
399. A
400. D
401. B
402. C
403. B
404. C
405. A

406.	D	451.	A	496.	A	541.	D	586.	B
407.	B	452.	D	497.	A	542.	D	587.	A
408.	A	453.	A	498.	C	543.	C	588.	A
409.	D	454.	A	499.	A	544.	A	589.	A
410.	B	455.	A	500.	A	545.	D	590.	D
411.	D	456.	C	501.	D	546.	D	591.	A
412.	A	457.	B	502.	B	547.	D	592.	D
413.	D	458.	A	503.	C	548.	C	593.	C
414.	D	459.	A	504.	C	549.	A	594.	D
415.	A	460.	B	505.	C	550.	B	595.	D
416.	B	461.	B	506.	D	551.	A	596.	D
417.	D	462.	A	507.	D	552.	D	597.	A
418.	D	463.	C	508.	D	553.	A	598.	B
419.	D	464.	A	509.	A	554.	A	599.	C
420.	A	465.	A	510.	B	555.	A	600.	D
421.	B	466.	C	511.	D	556.	C	601.	B
422.	D	467.	B	512.	D	557.	A	602.	C
423.	C	468.	C	513.	D	558.	D	603.	A
424.	C	469.	A	514.	C	559.	A	604.	A
425.	B	470.	C	515.	C	560.	A	605.	B
426.	C	471.	A	516.	D	561.	A	606.	A
427.	D	472.	A	517.	D	562.	B	607.	B
428.	D	473.	D	518.	D	563.	D	608.	C
429.	C	474.	A	519.	D	564.	B	609.	B
430.	A	475.	C	520.	D	565.	D	610.	A
431.	D	476.	C	521.	B	566.	A	611.	D
432.	D	477.	A	522.	D	567.	A	612.	B
433.	A	478.	A	523.	D	568.	C	613.	B
434.	C	479.	C	524.	D	569.	B	614.	C
435.	A	480.	D	525.	D	570.	B	615.	B
436.	A	481.	A	526.	C	571.	B	616.	A
437.	D	482.	A	527.	A	572.	A	617.	B
438.	A	483.	D	528.	C	573.	D	618.	C
439.	B	484.	A	529.	D	574.	A	619.	A
440.	D	485.	B	530.	D	575.	D	620.	C
441.	D	486.	D	531.	C	576.	D	621.	D
442.	C	487.	C	532.	D	577.	C	622.	D
443.	C	488.	A	533.	A	578.	D	623.	A
444.	B	489.	B	534.	A	579.	B	624.	B
445.	A	490.	D	535.	B	580.	C	625.	D
446.	B	491.	B	536.	D	581.	C	626.	C
447.	C	492.	B	537.	A	582.	B	627.	A
448.	D	493.	A	538.	D	583.	A	628.	B
449.	A	494.	A	539.	A	584.	C	629.	A
450.	D	495.	D	540.	B	585.	D	630.	D

631.	D	656.	A	681.	D	706.	B	731.	B
632.	D	657.	B	682.	D	707.	C	732.	D
633.	D	658.	C	683.	A	708.	D	733.	D
634.	A	659.	B	684.	D	709.	C	734.	D
635.	C	660.	A	685.	D	710.	A	735.	A
636.	B	661.	D	686.	D	711.	A	736.	A
637.	D	662.	B	687.	C	712.	C	737.	A
638.	D	663.	C	688.	B	713.	C	738.	C
639.	D	664.	A	689.	A	714.	B	739.	B
640.	B	665.	D	690.	B	715.	C	740.	D
641.	B	666.	B	691.	D	716.	D	741.	B
642.	C	667.	D	692.	D	717.	D	742.	B
643.	A	668.	D	693.	D	718.	B	743.	C
644.	C	669.	A	694.	C	719.	C	744.	A
645.	A	670.	C	695.	C	720.	C	745.	A
646.	D	671.	B	696.	A	721.	A	746.	D
647.	C	672.	D	697.	C	722.	D	747.	A
648.	A	673.	C	698.	B	723.	D	748.	C
649.	A	674.	C	699.	D	724.	A	749.	A
650.	B	675.	A	700.	A	725.	D	750.	D
651.	C	676.	B	701.	D	726.	C	751.	A
652.	D	677.	C	702.	B	727.	D	752.	D
653.	C	678.	D	703.	D	728.	C	753.	B
654.	B	679.	D	704.	D	729.	A	754.	A
655.	C	680.	A	705.	A	730.	B	755.	C

Match the Appropriate Option

1.	1	ii	2	i	3	v	4	iv	5	iii		
2.	1	i	2	ii	3	iii	4	iv	5	v		
3.	1	ii	2	i	3	iv	4	iii	5	v		
4.	1	i	2	ii	3	iii	4	iv	5	v		
5.	1	i	2	ii	3	iii	4	iv	5	v		
6.	1	ii	2	i	3	v	4	iii	5	iv		
7.	1	v	2	iv	3	iii	4	ii	5	i		
8.	1	v	2	iv	3	iii	4	ii	5	i		
9.	1	v	2	iv	3	iii	4	i	5	ii		
10.	1	v	2	iv	3	i	4	ii	5	iii		
11.	1	iii	2	v	3	iv	4	vi	5	ii	6	i
12.	1	v	2	iv	3	i	4	ii	5	iii		
13.	1	v	2	iv	3	iii	4	ii	5	i		
14.	1	i	2	ii	3	iii	4	iv	5	v		
15.	1	i	2	ii	3	iii	4	iv	5	v		
16.	1	i	2	iii	3	ii	4	iv	5			

17.	1	ii	2	iii	3	v	4	iv	5	i
18.	1	ii	2	iv	3	v	4	iii	5	i
19.	1	iii	2	iv	3	v	4	i	5	ii
20.	1	iii	2	i	3	ii	4	v	5	iv
21.	1	v	2	iii	3	iv	4	i	5	ii
22.	1	iii	2	v	3	iv	4	i	5	ii
23.	1	i	2	ii	3	iii	4	iv	5	v
24.	1	I	2	ii	3	iii	4	iv	5	v
25.	1	v	2	iv	3	iii	4	ii	5	i
26.	1	i	2	ii	3	iii	4	iv	5	v